LOCAL LEADERSHIP FOR

SCIENCE EDUCATION REFORM

Ronald D. Anderson
University of Colorado

Harold Pratt
Jefferson County Public Schools

KENDALL/HUNT PUBLISHING COMPANY
4050 Westmark Drive Dubuque, Iowa 52002

ISBN 0-8403-9947-2

Printed in the United States of America
10 9 8 7 6 5 4 3 2

Contents

Preface

This book is written to help you, a school leader, improve the quality of science education in your school or school district. As a science department chairperson, a teacher motivated to making changes occur in your school, a principal or vice principal wanting to help a science department make some improvements, a district science coordinator, or an assistant superintendent responsible for curriculum and instruction, you are in a position to foster such change.

Our basic premise is that meaningful change in science education programs will not happen without substantial initiative and effort at the local level. This statement is not at odds with the extensive state initiatives for educational reform, such as the development of standards and frameworks, and the development of national science standards or other national science education reform efforts. The premise has a strong scholarly foundation. Much research e.g., (McLaughlin, 1987; Crandall, 1982; Pratt, 1984; Anderson, 1990; Anderson, et al., 1994) has identified the importance of concurrent local school initiative and implementation if federal, state, or district programs are to have a lasting impact. Teachers and local administrators must make a major effort if they expect to have improved science programs. Federal and state efforts may help enhance the recipe, but the local initiative is as essential as flour for bread.

It is important to say a word about improvement. We do not believe there is a "crisis" in science education, caused by poor teachers and poor curriculum. As a profession, we have learned much in the last 30 years and made many improvements in our curricula and the way we teach students, but there is much more to do. The improvement of the science education program should be the ongoing goal of every teacher and administrator. Much of the world, including those countries we often consider as "third world", is putting a major effort into improving the educational level of their population in order to compete in the international market place. There is evidence that quality of life in this country is not what it could be because of our reduced ability to compete i
international business arena and inequities within ou

Improvement in the quality of our science education programs and the subsequent gain in student achievement are important for our society at large.

Since it is our intent to address these matters very directly, it seems appropriate to identify some of the origins of our perspective. Our background includes both practical experience and research. One of us (Pratt) has twenty-eight years experience as the science coordinator, and curriculum director for a large school system, in which extensive efforts at curricular and instructional change have been made over the years. For example, an activity-based elementary science program initiated in the late 60's has been sustained throughout the district ever since, with periodic review and revision, along with extensive staff development and other forms of continuing, systematic support for teachers. More recently, the district has developed its own middle school science curriculum materials, including student text materials, when the desired materials were not commercially available. The district life science program, which was developed with the support of National Science Foundation funds, was published commercially by a major publishing company. Although equal success would not be claimed for all of these efforts, this extensive experience is reflected in what follows. In addition, both authors have served as consultants and members of review teams in a number of different school systems of widely varied size. Recently Pratt has been a staff member of the team at the National Academy of Science, developing the national science standards.

As valuable as generalizations drawn from personal experience are, they must be coupled with those based on extensive and systematically controlled experiences. Empirical research has produced an impressive body of knowledge about teaching and learning science and about the processes of educational change.

The other author (Anderson) has been involved in educational research for about three decades, with particular attention to studies of change in science education practice. As part of this research work, considerable effort has been devoted to consolidating the results of the extensive science education research literature. For example, much of the quantitative research in science education was examined through meta-analysis (Anderson, et al., 1983), cost-effectiveness analysis was applied to research on the effects of various science education improvement practices (Anderson, 1990), and an extensive review of the research on science curriculum reform has been prepared (Anderson, et al., 1994). Recently, under the sponsorship of the U.S. Department of Education, he has been directing a series of case studies of schools across the country which are undertaking science, mathematics or general curriculum reform.

Through the results of meta-analyses and other reviews of research such as those mentioned above, important generalizations are available about the effectiveness of innovations, varied curricular approaches, and a variety of educational practices. No action should be assumed good or bad without considering what research may have to say about it. In the last two decades the process of educational change, the role of school leaders, and the characteristics of effective schools have become important areas of research that are very useful to school leaders who are initiating change in science education. In particular, current information about the reality of school practice, the values and concerns of teachers, and the culture of schools is growing as a result of science education researchers' attention to case studies and other forms of qualitative research.

The plethora of generalizations available from research, along with the recommendations of various national groups, only increases the dilemma of choice. If you are initiating a change you have many decisions to make. In addition to the National Science Education Standards, the work of The American Association for the Advancement of Science in Project 2061, and the examples of various schools in the National Science Teachers Association's Scope, Sequence and Coordination project, a potential source of assistance for decision makers comes from various forms of policy research. In particular, attention will be given to a recent cost-effectiveness analysis of a large number of possible actions (Anderson, 1990). This analysis, in which both of us participated, considered the tradeoffs among costs and potential benefits of initiating one or more of various reform actions.

The purpose of this book is to help with the process of improving your science program. It is not a review of what programs to pick or what actions to take, but we will elaborate a process and help you consider how to best proceed. Good luck; it's an important task that needs to start in every school and/or district across the country. We cannot afford not to begin.

Ronald D. Anderson
Harold Pratt

References

Anderson, R.D., et al. (1983) Science education: A meta analysis of major questions. *Journal of Research in Science Teaching*. *20* (399).

Anderson, R.D. (1990) Policy decisions on improving science education: A cost-effectiveness analysis. *Journal of Research in Science Teaching*, 27(6), 553-574.

Anderson, R.D., et al. (1994). *Issues of Curriculum Reform in Science, Mathematics and Higher Order Thinking Across the Disciplines*. Washington, DC: U.S. Department of Education.

Crandall, D. (1982) *People, Policies, and Practices: Examining the Changing School Improvement* (ten volumes). Andover, MA: The NETWORK.

McLaughlin, M.W. (1987) Learning from experience: Lessons from policy implementation. *Educational Evaluation and Policy Analysis*.

Pratt, H. (1984) Science leadership at the local level: The bottom line. *NSTA Yearbook: Redesigning Science and Technology Education*. Washington, DC: National Science Teachers Association.

1

Life on the Firing Line

"National Standards Will Increase Expectations for Schools and Students"

"Crisis in Science Education Continues"

"National Education Goals Call For U.S. To Be First in Science and Math Achievement by 2000"

"National Science Standards Will Set Pattern and Vision for Reform"

The headlines in the special edition of the weekly magazine get your attention immediately. You skim down through the articles and read of the need for national standards to improve the quality of education that students need to succeed in the 21st century. The articles address the need to overcome problems of declining test scores, the absence of science in the elementary schools, negative comparisons with other countries, and the negative student attitudes towards science. Although the tone of the articles is often critical, the message is one of thoughtful optimism; the national standards and the support they are receiving seem to provide hope for reform. There is much in the articles that makes you wonder about the quality of the science program in your school and question how close it is to the recommendations in the standards.

Sidebar 1.1

NATIONAL EDUCATION GOALS

GOAL 1: By the year 2000, all children in America will start school ready to learn.

GOAL 2: By the year 2000, the high school graduation rate will increase to at least 90 percent.

GOAL 3: By the year 2000, American students will leave grades four, eight and twelve having demonstrated competency in challenging subject matter including English, mathematics, science, history, and geography; and every school in America will ensure that all students learn to use their minds well so they may be prepared for responsible citizenship, further learning, and productive employment in our modern economy.

GOAL 4: By the year 2000, U.S. students will be first in the world in science and mathematics achievement.

GOAL 5: By the year 2000, every adult American will be literate and will possess the knowledge and skills necessary to compete in a global economy and exercise the rights and responsibilities of citizenship.

GOAL 6: By the year 2000, every school in America will be free of drugs, violence, and the unauthorized presence of firearms and alcohol, and will offer a disciplined environment conducive to learning.

GOAL 7: By the year 2000, the nation's teaching force will have access to programs for the continued improvement of their professional skills and the opportunity to acquire the knowledge and skills needed to instruct all American students for the next century.

GOAL 8: By the year 2000, every school will promote partnerships that will increase parental involvement and participation in promoting the social, emotional, and academic growth of children.

The message of the standards appears to be well thought out, albeit somewhat idealistic and maybe a little naive. "Even so, maybe I should read this a little more carefully," and as you do, it becomes clear that the recommendations from the standards are well documented, logically presented, and do provide direction and vision for moving into the next century. "There must be something to this; I wonder how our schools would measure up? I wonder if our staff would agree with the recommendations?"

If the person in the above scenario is your superintendent and you are the principal of the local high school, or an elementary school principal, the science supervisor, or the science department chairperson in your school, the next step probably is a phone call or note from the superintendent asking you to stop by and see him or her in the next day or so. Once you get to the office, the superintendent says, "Have you seen this article? I just read it, and I'm wondering how our schools look in science. We ought to take the standards seriously. I also got a call from a school board member last night asking me the same question, and this morning in the office I received a call from one of the prominent citizen leaders of our district. They are saying we ought to do something about the quality of science instruction in our schools."

The superintendent then begins to get down to business and provides some support, directions, time lines, and desirable outcomes; but leaves the process up to you. In essence, the message to you is: "You will be given support in terms of some release time for you and the members in the department to find out what needs to be done to improve the quality of science instruction our students are receiving and to bring back a written report in the next six months. Next, you will have to present the report to the school board. It should contain recommendations, costs, a clear description of the outcomes, and some evaluation procedures to assure the school board that the results can be obtained if they provide the necessary financial support."

This process started from the top down, as so often is the case. But consider another scenario: You may be a member of a team in an elementary school or a secondary school science department in your district. In this instance you make copies of the article and circulate them to other members of your team or department, and

Sidebar 1.2

VISION FOR SCIENCE EDUCATION FROM NATIONAL SCIENCE EDUCATION STANDARDS

- **All** Students should have the opportunity to learn science;
- With appropriate opportunities, all students can learn science;
- Students should learn science in ways that reflect the modes of inquiry that scientists use to understand the natural world;
- Learning is an active process that occurs best when students act as individuals who are members of a community of learners;
- The quantity of factual science knowledge that all students are expected to learn needs to be reduced so that students can develop a deeper understanding of science; and
- Teaching and assessment need to be considered in context and relationship to each other.

at the first opportunity in a team meeting you raise the question, "Does this apply to our school?" "What are the implications of the national standards for our school?" Almost immediately one or two other members of the department say, "No, that doesn't apply to us, those data really come from schools where they don't spend as much to support education as they do here in our community." But other members counter with, "How do we know, we haven't taken a good look at science test scores in several years? We haven't used the standards as criteria to rate the quality of our program." or "Our elementary program really hasn't changed much in the last seven or

---

From National Research Council. (1993) *National Science Education Standards: July *1993 Progress Report*, Washington, D.C.

eight years, and we are not doing nearly as many hands-on activities as we really could. Do we really know the attitudes of our students towards science? Is it as good as we think it is?"

So almost as though it had been planned that way, the department finds itself in a spirited but highly professional, even sometimes opinionated, discussion about the quality of science in their school. After several meetings and discussions on this topic, informal discussions in the hall between periods, and a conference with the principal, the group decides you really ought to look into this situation to see if it has something to do with your local program. The principal has indicated he/she will find some support to give you an opportunity to do a study and develop some recommendations. If necessary, you will even be encouraged to go on to the superintendent and the school board, but other than that he provides no particularly detailed suggestions or guidance on how to proceed.

What do we do first? How do we go about this process of self-evaluation of the quality of science in our school?

Both scenarios are possible beginning points for the type of activities to be addressed in this publication. In both cases there is an apparent commitment to taking time for careful thought, analysis, and planning. There is the possibility of some resources being allocated which would make the planning worthwhile. There is a commitment to having good science education in the schools and an expectation that if it is not already present, steps should be taken to bring it about.

We are going to build on the commitment to improve or reform the local science program in this book, a commitment to systemically and systematically plan and implement a long term process to improve the quality of science education in a school or district. It also should be recognized that there are other scenarios which occur—scenarios which are not based on long-term plans and carefully considered actions, nor the availability of the resources needed to get positive change.

Sidebar 1.3

ESTABLISHING THE IMPROVEMENT PROJECT

A very useful monograph, *An Action Guide to School Improvement* by Loucks-Horsley and Hergert describes five tasks necessary to establish the project and get started.

1. Clarify your Charge
   What are the desired outcomes of the project?
   What is your role in reaching them?

2. Negotiate for Resources
   Time
   Money
   Services

3. Build a Base of Relationships
   Support for the project
   Positions or opinions of key people
   Communication with key people before starting major activities.

4. Consider Using an Outside Consultant
   Advantages
      Provision of time and effort often not available to staff
      Balance the power between factions
   Disadvantages
      Cost
      Can be dangerous if they impose their ideas

5. Form an Improvement Project Team
      A small team (5-15 people) representing the various constituents of the school or district
      Team's role and authority clearly defined

From S. Loucks-Horsley & L. Hergert (1985) *An Action Guide to School Improvement*, Andover, MA: ASCD and The NETWORK

For example, the assistant superintendent of a local district telephones the science education specialist at the state university. "We want to do something about our elementary science program. Our budget this year has money for a new textbook adoption, and a committee has recommended to the board that we adopt the new edition of the one we have been using for some time. But we want to do more than just get the most recent materials. We were wondering if you would be available to do an after-school workshop for our elementary teachers on three different days during the fall. With the freshness of the new edition and the enthusiasm generated by these hands-on workshops, we are really hopeful that our test scores in science will go up next year."

The inadequacies reflected in the last scenario will be highlighted when compared to the following assumptions reflected in this book.

Some Assumptions

No improvement can take place without a change. Change per se is not being advocated but if improvement is to occur, so must change. Therefore, it is important to set down some of the assumptions or beliefs that undergird our presentation. They include:

1. Significant change is possible. The new curricular programs of a generation ago did improve a variety of student outcomes. While research shows that innovations sought in school settings usually were only implemented at a low level, research also shows that substantial change is highly probable *if* certain elements are included in the process.
2. Significant change is not easy. It is a complex process that requires substantial commitment. Significant change requires a long term commitment, three to five years typically are needed to produce lasting change of any consequence.
3. Leadership is essential. Research of the last decade or so has shown how important the leadership of principals is, but they are not the only sources of the needed leadership. Central school district personnel and key leaders within the staff of a building are important sources as well.

4. Changing the roles or behavior of people is the most difficult aspect of change. Changing organizational structure or programs is relatively easy. But if the changes require different behaviors for personnel, e.g., for teachers, the task at hand is much more formidable. On the other hand, such changes in behavior probably are the most critical aspect of the educational changes being sought.
5. The process itself is critical. The person seeking change can not just focus on the ends. The means of reaching these ends are crucial.
6. Finally, and probably most important: change must be made on a *systemic* basis. In other words, the entire system must be affected. No one change should be initiated with high hopes of improvement. Several actions are probably needed, and they must be approached in a coordinated manner. In many ways this latter point is a major theme of the entire book.

Need for Systemic Change

The words "reform" and "restructure" fill the educational literature today. The implication of this language is that the entire system must be changed. We must do business very differently than we have in the past. It also means we can't just change science; all subject areas must be improved. We accept this premise, but will deal here with the science portion of this reform.

Think about the following components of the educational "system":

Goals

Curriculum

Staff Development

Student Assessment

Program Assessment

Budget and Other Administrative Support

A change in curriculum without rethinking the goals of the program will likely lead to major inconsistencies and confusion. If staff development is not available to support the new curriculum, it will probably go unused by many and misused by others. In like manner, student testing and evaluation must be consistent with the curriculum activities and goals, or the results will be meaningless, if not counter productive.

There is another dimension to the system, namely the state and local hierarchy. Schools in some districts operate quite autonomously; others are very much a part of a political system that may stretch through the local administration, to an intermediate agency, on to the state level. If you are in a school where support (and often directives) come from the central office of your district or from the state level, it is important to find out what goals, programs, and frameworks are available before you take a direction of your own.

The point should be clear. An improvement project that involves only part of the system will be far less effective than one that is truly systemic.

Phases of an Improvement Plan

Just as there are many parts of the system to consider, there are a number of varying steps or phases which underlie and organize the message of this book; they generally come in the approximate order given here:

1. *Goal Setting* — The goals of science education have been well articulated in the National Science Education Standards and reinforced in the Benchmarks and Science for All Americans, published by the American Association for the Advancement of Science in Project 2061. Chapters 2 and 3 will focus on the matter of transforming these goals into local action, a high priority that should direct the following activities.
2. *Needs Assessment* — This term is used in a broad sense and has many forms; it is not just a simple process of distributing and tabulating a questionnaire. The details of this process will be outlined in Chapter 3.

3. *Development* — The desired goals must be embodied in some product such as a set of curriculum materials, teaching strategies, or some combination. Chapters 4 and 5 will address these important strategies for improvement.
4. *Implementation* — After planning the work, it is time to work the plan. Materials must be acquired, staff development carried out, and all of the other actions taken to insure that change actually occurs and is put into appropriate use. Chapter 6 will discuss both research on effective implementation and the practical steps that need to be taken.
5. *Maintenance* — The initial implementation program is not enough. Steps must be taken over a period of years—and possibly indefinitely—to support and sustain the changes and insure that they remain an integral part of the school or school system. The long term nature of implementation and the maintenance necessary to institutionalize improvement also will be discussed in Chapter 6.
6. *Evaluation* — Program evaluation starts near the beginning of the improvement process and is used—among other functions—to guide and shape the total process outlined in steps 3, 4 and 5. Assessment of student achievement becomes an important part of the evaluation plan as the innovations become well implemented. Chapter 7 will address these important, and often overlooked, activities.

In Chapter 8 we will take a careful look at the leadership needed to get the job done. Improvement is a people process, not just a series of mechanical steps. Finally, in Chapter 9 we will summarize many of the important points of the book by pointing out some of the pitfalls of working through the improvement process.

Reform Revisited?

But haven't we tried to reform science education before? And if it didn't work in the 60's and 70's, how will the "new reform" movement based on national standards, better assessment, and a new round of national curriculum products be any more successful this time? Why are we still in such a fix? Actually, very few schools have tried a systemic reform effort before. The curriculum reform movement that exploded on the U.S. science education scene during the post-Sputnik era included some elements of what is being advocated here, but rare indeed is the school or district in which a systemic approach has been seen in actual practice. The experience and research of the last 25–30 years has taught us much. These lessons, coupled with the newer, more systemic perspective in the national standards, and an emphasis on local leadership, form the foundation of this book, support in science education.

2

What Do We Really Want Our School Science To Be?

You may wonder why the news story on declining science scores so fully caught people's attention in the previous chapter's opening scenario. Even if the obvious political explanation is accepted, a slight variation of the question still faces us, "Why does the public have such an interest in this matter?"

The answer is that school matters; there is a conviction on the part of both the public and educators that education is important and that schools could be doing an even better job of providing this education. Underlying this perspective are a number of assumptions including the following:

1. As already indicated, schools matter.
2. An understanding of science is crucial to the well being of all individuals.
3. An understanding of science on the part of its citizens is crucial for the economic and social well being of our nation.
4. Test scores are an important indicator of the learning that students are gaining.
5. The schools could be doing a better job of educating students in science.
6. We know what science education ideally should be.

Do We Know What It Should Be?

This last assumption—that we know what science education *should* be—deserves considerable attention because when examined closely the matter is far more complex than generally assumed. A diversity of opinions among professionals and the public means that consensus on these matters is not automatic. There are varying degrees of understanding of the substantial amount of empirical research about teaching and learning having implications for one's ideal of what science education should be, as well as varying value judgments about personal, societal and educational issues.

In addition to its complexity, the question of what science education ideally should be is of crucial importance. Knowing what facets of science will be emphasized, and why, is crucial to those curricular decisions. Knowing how students are expected to be different as a result of their science education is central to deciding how science will be taught and what experiences will be provided in the classroom. One's answers to these questions are the foundation of a coherent rationale for science education reform.

The variations in perspective are large, the range of alternatives in educational practice is equally large and developing a consensus of these matters by a group of educators is no simple task. But without this common understanding, the potential for making significant improvement in educational practice is sharply reduced. The answer to the question of what science education should be, is not as obvious as usually assumed.

Is a Needs Assessment the Source of Our Answer?

A needs assessment may provide direction and clear goals and produce the consensus judgments needed for action. Unless such a needs assessment is clearly defined, however, it will not provide the needed answers. Simply developing a questionnaire, having it completed by the appropriate people and tabulating the results will not do the job.

The process begins with identifying alternative visions of what this ideal could be. Such visions should be based on careful analysis of alternatives in the literature and application of the results to your particular situation. The new National Science Education Standards and the recommendations of such groups as AAAS's Project 2061 deserve careful study and exploration of their applicability to your local situation.

What currently exists in school science is not necessarily what we want. But if it is not, what *do* we want?

In addition to these science education publications, the publications on general educational reform make a contribution. The question of goals has a prominent place in essentially all of these documents. So, before turning to the process of conducting a needs assessment in a later chapter, attention must first be given to this careful analysis and to discussion of alternative conceptions of science programs.

What we are presenting here is no substitute for reading such a landmark document as the National Standards. As you reflect on the Standards, however—and how you can use the information in this book to help you implement them—we suggest that you think about how your new ideal of science education will exhibit itself in practice in your local situation. School personnel at the local level must develop their own personal convictions about what they want science education in their schools to be. Without such convictions and some local consensus on these matters, the prospect of having a major impact is minimal.

To assist in this reflection process, an analysis framework is presented from analyses of the many actions being recommended to change science education today. This new vision of science education, what you may call your "desired state" of science education will emerge from some combination of actions you take to change current practice. This framework is presented in three dimensions:

(a) *quantity* (more science),

(b) *quality* (better instruction), and

(c) *appropriateness* (more suitable science).

Essentially, all the specific actions recommended for changing science today are expected to produce improvements in one or more of these three dimensions, as discussed below.

Quantity: More Science

A popular contention today, one we share, is that students should be getting more education in science than they have been. For example, senior high students should take more than one science course—the common minimum in most regions until the mid-eighties. This particular action—requiring more science to graduate from high school—should be examined briefly in terms of probable costs and effectiveness.

The apparently simple action of increasing high school graduation requirements in science deserves careful scrutiny—as do most apparently simple solutions to problems. It should be recognized from the outset that it mostly affects non-college bound students, rather than the typical college-bound students who already were taking more than the minimum amount of science.

Some may question why these "non-academic" students need more science, in effect displacing other work from their programs, such as vocational courses. To those of us with a strong interest in science the answer may seem obvious, but others may need to be reminded of the rapidly increasing influence of science and technology on our world and the many science-related issues that must be understood for intelligent voting, personal decisions, and simple appreciation of many dimensions of life. It must be recognized, however, that there is a real cost to students' taking more science, namely the loss of instruction in other areas, such as fine arts, music, and technical education, which the student now will be unable to take.

But assuming a consensus that more science is desirable, there is still the question of whether or not an increased graduation requirement is the way to achieve this goal. There are other possible approaches to increasing the time engaged in learning science, such as increasing the length of the school day or the school year, increasing the proportion of total class time devoted to instruction, and increasing the amount of homework. Whatever the means by which change comes about, a case can be made for extensive science instruction in high schools for *all* students. But more science for all does not necessarily mean more of the same thing that is already in the curriculum, a matter to which we will return in our next two dimensions.

Having begun this discussion of more science with the high school level does not imply diminished importance for the issue in the elementary schools. There is extensive evidence—both in surveys of school practice and case studies conducted in school classrooms—that the amount of science taught in elementary schools is quite limited. We need more. But it is not just that we need more; we need better science instruction, the matter to which we turn attention next.

Quality: Better Instruction

A second dimension of the desired state of science education is greater quality, i.e., better instruction. What is it we are seeking in terms of greater quality instruction? The following may serve as a starting point for discussion of this matter:

1. More interesting and stimulating interaction with science on the part of students,
2. More up-to-date science, which reflects reasonably well the current developments in the given field of science,
3. More pedagogically sound knowledge, i.e., important concepts and their inter-relationships, rather than simply vocabulary and isolated facts, and
4. More accurate science, i.e., concepts that are accurate and full of enough meaning that they are understood in depth and in the context of the many inter-relationships of which they are a part.

How do we achieve these goals? The list of commonly suggested actions is long indeed; the following are representative:

- new standards for science teacher preparation programs,
- increasing teacher salaries,

- initiating merit pay for teachers,
- better teacher evaluation,
- reducing teacher work loads, i.e., fewer students and/or classes,
- workshops and other education for in-service teachers,
- teacher centers where teachers can work together on program development,
- extended year contracts for teachers for program development work,
- more and better materials and equipment, and
- providing summer employment for science teachers in scientific research.

But will such actions have the desired effect, i.e., higher quality science education? The answer in a nutshell is that any one of them, *as a single action,* is unlikely to have a substantial impact. This argument will be developed in some detail later; real hope lies in a *combination* of wisely chosen and appropriately initiated actions.

It also is obvious that one must consider the fundamental orientation of such actions as program development by teachers and in-service workshops conducted for their professional growth. If better science education is to result, the nature of the activities must be such that fundamental instructional principles are addressed. A general orientation to instruction that research has established as being of fundamental importance, is an outlook often labelled a "constructivist" approach. It is grounded in cognitive science research that shows learning is more than acquiring knowledge. Understanding requires items of information and a structure in which this information is organized (West & Pines, 1985, Shuell, 1985). This structure can not just be given to the student. It must be "constructed" by the student and the process of adding new science concepts to it at times is quite difficult.

The topic of constructivist learning is big, but it can be summarized adequately enough in the following four basic generalizations. First, learning is a process of students constructing their own meaning. Understanding, for example, that "cold is the absence of heat," is not a matter of remembering this six word phrase so it can be fed back to the teacher on a test, nor is it a phrase to be pulled up when one needs to think about heat and temperature. It is a conception that comes from understanding something about the nature of heat energy, and "trying out" this understanding in a variety of situations faced in everyday life until it is clear that they make sense to the learner. Without this process of constructing a new conception of "cold," the six-word phrase has no meaning and probably will soon disappear from memory even as a phrase.

Second, learning depends upon the preconceptions students bring to a subject, i.e., meanings they have already constructed at a prior time (Shuell, 1985; West & Pines, 1985; Linn, 1986). They begin with what they know and build additional knowledge from that foundation. They learn "by assimilating new concepts to old frameworks (a type of evolutionary learning) or by accommodating new frameworks from old ones (a type of revolutionary learning)" (Wittrock, 1985, p. 260). A major difficulty in science is that many fundamental concepts are counterintuitive and are inconsistent with what one learns from everyday experience. Newton's laws of motion, for example, are not intuitively obvious and, in fact, for most people are in conflict with their interpretations of everyday experiences. The students who say it is not plausible that a satellite is placed in orbit around the earth and then keeps circling the earth for years without some kind of engine "to keep it going" are thinking about the situation in a manner that may be quite consistent with everything they have experienced in everyday life. Students often come to the learning task with misconceptions that must be replaced—misconceptions that have been useful to them in their everyday experience. The relearning process is involved and takes time. It is not simply a matter of adding a new piece to their storehouse of knowledge.

Third, learning is dependent upon the context. "All relations are embedded in a larger context. . . . Thus, in order to understand one single relationship, one needs to understand a whole context." (West, 1985, p.194). "A single concept may mean one thing within one framework and something slightly different within another context." (Pines, 1985, p. 109).

Specific examples from various areas of the natural sciences may enhance this matter of context.

> For example, the chemistry student who acquires the meaning of entropy sees solubility, reaction rates, and osmosis with substantially new meanings. To

> see what such superordination can do in a discipline, think about what relativity has done for physics or plate tectonics for geology. (Pines, 1985, p. 194)

In a plant context, "life" has certain meaning; in an animal context it has somewhat different meaning. In a non-biological context the concept may mean something altogether different. Devoid of all context "life" is little more than four different letters of the alphabet in a particular order. Until the biological concept of life is explored in a variety of contexts, the learner probably will not understand it. The importance of such exploration and the difficulty of acquiring many basic concepts is illustrated by the results of one researcher who found that about one-fourth of the students in a seventh grade class believed that fire was alive (Bielenberg, 1993).

Fourth, meaning is socially constructed; understanding develops through interaction between student and teacher and between the student and other students. Students' construction of new understandings is difficult work and is greatly facilitated by interaction with others.

Good teaching fosters learning in which students can engage in the construction of meaning. It is not a matter of presenting information. It is a matter of helping students construct meaning for themselves that takes account of past conceptions and fits the relevant contexts. Current science teaching in the schools, however, is heavily focused on low level knowledge. This problem is compounded further by the fact that when more complex concepts are taught, they often are taught as if they were of the same character, i.e., as if they were simple pieces of information to store away. Effective science teaching must take account of the great variations in the character of the content and the individual situation each learner brings to the learning. Such teaching requires more than good communication skills and having good curriculum materials. It requires teaching competencies and initiative of the highest order.

From this constructivist perspective on learning and teaching, it is apparent that just having time for teachers to engage in program development and in-service education by itself is far from enough. Program development and in-service education must grapple with basic issues of instruction and deal with instructional materials accordingly. To have better science education requires that basic issues be addressed.

Appropriateness: Science that is More Suitable

The third dimension of our desired state of science education brings us to the basic question: what knowledge is of most worth, or what should be the content of the curriculum? It may well be the most important of the three dimensions.

The content chosen is a reflection of the goals one has for science education; describing the desired content will serve to communicate in a tangible way something of the purpose of teaching science. There are many ways in which such curriculum content has been categorized and communicated. We have chosen to use here the following four categories: (1) science as inquiry, (2) science subject matter, (3) scientific connections, and (4) science and human affairs. These categories, of course, imply nothing about how the content will be organized in the curriculum. They are goals of instruction that will be organized in different ways within different curriculum programs—understandings that will be acquired through a wide range of different instructional experiences. The four categories of science content are elaborated below.

1. *Science as inquiry*. Inquiry is the multi-faceted means by which scientists pose questions and seek the answers to them. Individuals should have an understanding of the processes or modes of inquiry by which the body of scientific knowledge is acquired.
2. *Science subject matter*. Each student should have an understanding of major scientific ideas from the various scientific disciplines. Even though the focus is on fundamental concepts and principles rather than facts and vocabulary, the number of concepts a student can learn is necessarily limited; selections must be made and attention given to those concepts that are foundational for understanding science and applying it.
3. *Scientific connections*. Science does not exist in isolation; its components within and among the various disciplines are connected—just as science is connected with mathematics,

technology, and engineering. These connections are part of the content of the curriculum.

4. *Science and Human Affairs.* Both the inquiry and subject matter of science have many applications to the personal lives of students, to problem-solving and decision-making, and to the resolution of many societal issues. The importance of these applications continually grows in our increasingly technological world. The relationship between science and human affairs extends beyond such applications in that science and technology on the one hand and the culture and context in which they exist on the other hand, influence and affect each other in many ways.

Most observers would agree that all of these categories deserve attention in the curriculum; the issue is how much attention each should receive. As will be described in some detail in the next chapter, it appears that current *practice* in the schools—but not necessarily what is advocated or what professionals believe is being practiced—generally is to devote at least 95% of instructional time to science subject matter (facts, concepts, theories and themes) independent of the scientific inquiry by which this knowledge was acquired or the applications of the science subject matter. In other words, number 2 of the four categories given above gets more than 95% of the attention.

Our position is that the current state of science education as generally found in the schools is not the desired state. The processes of science and the applications of science knowledge need much more than the minimal attention they currently receive. Even the study of fundamental science knowledge would benefit from being addressed more frequently *in the context of the inquiry processes and applications of science.* In such a context it can be expected to be less sterile and irrelevant than the current situation where the focus so often is on vocabulary, isolated facts, and truncated concepts. A different grid for selecting ideas to study will result when the goal is more than simply preparing for the next level of schooling. Furthermore, this attention to the applications of science knowledge results in increased motivation and provides another context in which to reinforce the knowledge learned.

Attention to the scientific connections described above—i.e., the interdisciplinary aspects of science and the connections of science to mathematics, technology, and engineering—has important benefits as well. It fosters motivation and reinforcement and gives students a better sense of the interconnections in science. Its importance is recognized, for example, in the work of two recent major projects: Project 2061 at the American Association for the Advancement of Science (1989) and the Scope, Sequence and Coordination Project of the National Science Teachers Association (1989).

The argument for addressing inquiry and the processes of science has been developed so extensively in the last three decades that it seems redundant to do more with it here. It may be helpful to note, however, that the arguments presented often have dealt not only with the processes of science as content in the curriculum, but with process as a teaching strategy or manner of student learning. While not denigrating process used in these latter senses, the aspect being addressed here is study of the nature of science as curriculum content. We find convincing the arguments that an understanding of the processes of science is important for an appreciation of science itself, as well as for its application to personal needs, societal issues, and decision-making.

The argument for attending to the applications of science is grounded in our understanding of the transfer of learning. Knowledge learned in the abstract in conventional science classes is not automatically transferred to addressing personal needs or societal issues. Students learn how to apply science knowledge in such contexts by practicing it, and in our increasingly scientific and technological society, the need for these competencies continues to grow (for numerous examples of specific applications see the report of Project Synthesis, i.e., Harms, et al., 1981).

Before leaving this discussion about making the curriculum content more appropriate, attention must be given to the most frequent argument against change: there is not enough room in the curriculum; students need all the available time to study the material they will need when they get to college. The pervasive presence of this "preparation ethic" among teachers leads to the question of the importance of this preparation. How critical is one's background to success at the next level of schooling? The available empirical evidence seems to indicate that its importance is blown out of proportion. In a review of studies of the correlation between background in high school chemistry and success in college chemistry, Ogden (1976) found a very low relationship. Medical schools, which have changed their entrance requirements to admit students both with and without strong science backgrounds have found that they do equally well in medical school (Woodward, 1983). More recent studies showed little evidence that taking high school chemistry or physics courses improved student grades in college science courses (Yager & Krajcik, 1989; Yager, et al., 1988). The obsession with "covering" material, which is so much a part of the preparation ethic, should not be allowed to detract from efforts to improve the curriculum. More important than coverage is for students to be interested in science, understand the nature of the scientific enterprise which produces the knowledge, understand its relevance to personal needs and pressing societal issues, and most important of all, acquire a true understanding of those concepts which are addressed. The constructivist learning principles discussed earlier lead to encouraging the idea of "less is more," i.e., being selective about what is covered, but covering the selected content in such a manner that the concepts are truly understood.

Given the enormity of the change suggested when advocating an increase in the percentage of time devoted to process and applications, it may be of interest to know the position taken by others. For example, 23 science education leaders participating in Project Synthesis analyzed the results of three national studies of science education and culminated their work with strong advocacy for the inclusion of substantially more applications in the curriculum (Harms, et al., 1981). In a survey of leading teachers, legislators, school board members, university professors and administrators, school administrators, and vocational educators in the state of Colorado, approximately two-thirds of the people expressed support for the shift to substantially more applications in the science curriculum (differences among the various subgroups were not substantial). The reasons presented by the people favoring this shift resembled many of the arguments expressed above. Those who did not want significant change from current practice in this regard feared that the curriculum would be "diluted" and subject matter not emphasized or that there would not be enough emphasis, on "discipline" as a goal of instruction. Although there is not a clear public consensus on this issue, it is apparent that there is strong support for *at least* experimenting with some alternatives and finding out what they would look like in practice.

Sidebar 2.1

A NATIONAL SCIENCE TEACHERS ASSOCIATION POSITION STATEMENT

Recommendations for K-12 Grade Levels

Time on science learning

- ▼ Lower elementary level (grades K-3): a minimum of 1 1/2 hours/week of science should be required.
- ▼ Upper elementary level (grades 4-6): a minimum of 2 1/2 hours/week of science should be required.
- ▼ Middle/junior high school level (grades 7-9): a minimum of 1 hour/day, for at least 2 full years of science, should be required of all students.
- ▼ Senior high school level (grades 10-12): a minimum of 1 hour per day for 2 full years of science, should be required. The courses should represent a balance of physical and life science.

More recently, the report of Project 2061, with the full endorsement of the board of its sponsoring agency, the American Association for the Advancement of Science, advocated similar goals, with particular attention to covering less and teaching more thoroughly what is retained. There is a consensus among leaders in the community of scientists that a new goal orientation is needed.

The importance of the issue at hand is even more apparent when one realizes that the move in the 1980's to higher graduation

---

From *An NSTA Position Statement. Science-Technology-Society: Science Education for the 1980s* (p.3) by National Science Teachers Association, 1982, Washington, DC: NSTA.

requirements has placed large numbers of "general" students into a very abstract curriculum developed with college-bound students in mind. The need for more attention to applications and processes of science in the curriculum may be more obvious when less interested students are faced with a requirement that they take additional science they wish they could avoid. Although the issue may be especially obvious when addressing the needs of the "general" student, there is no intention of asserting here that college-bound students necessarily should give less attention than other students to the applications of science.

Zeroing In On *Your* Desired State

Having described the quantity, quality, and appropriateness dimensions of a sound science program, it is time to zero in more specifically on just what the desired state of a science program is. In doing so, attention will be directed to (1) key matters and (2) pressing issues within each of these three dimensions. The key elements of a strong science education program are those crucial elements which should be in place if a program can be expected to provide quality teaching and learning in science. The second category refers to issues or questions local science education personnel must address in creating their own personal vision of what they want science education to be in their setting. The former are targets for leaders seeking a quality science program; the latter are issues they must address in getting these elements appropriately in place in their local context. Although the discussion will cut across grade levels, specific attention will be given to selected implications at the three major educational levels: elementary, middle/junior high, and senior high school.

Quantity

Key Elements. Given the considerable variation in the amount of science taught in U.S. elementary schools—and the generally minimal amount in most cases—the need for science to be taught throughout the junior high/middle school years, and the need for sufficient science instruction for students going in a variety of directions after high school, the quantity of science available at all levels is an important dimension. The following key elements of a school district science program are recommended.

- Elementary school students get a significant amount of science instruction.
- Science is a part of each student's program throughout the middle/junior high school years.
- All high school students receive sufficient science education to make them scientifically literate.
- High school students going directly into occupations receive sufficient education in science for employment in today's technological economy.
- Students pursuing higher education should receive sufficient science and mathematics education in high school to enable them to pursue the majors of their choice.

Issues. Given the many topics competing for a place in the curriculum, it is not surprising that at all levels the major issues revolve around choices that must be made among various content areas. It is apparent that any move to give science its rightful place will require addressing the question of how big its place is.

How do we find time in the elementary school day for science, i.e., what do we give up in the existing curriculum to gain this time?

How can science be maintained as a *continuous* course of study in the middle/junior high school curriculum in the face of competition for space from other courses?

In the senior high school, how much science should be taken by the general student and by the college preparatory student?

Sidebar 2.2

Six criteria for content selection and curriculum design are suggested.

- they build upon children's prior experiences and knowledge;
- they capture children's interest;
- they are interdisciplinary so that children see that reading, writing, mathematics, and other curricular areas are part of science and technology;
- they integrate several disciplines;
- they are vehicles for teaching major organizing concepts, attitudes, and skills; and
- they allow a balance of science and technological activities.

Quality

Key Elements. The key elements of quality involve both the curriculum and the form of instruction. The following, each of which has been promoted in many national studies and reports, constitute our list of the major elements of quality science education.

- Less is more: fewer topics are studied in some depth—the focus is important concepts, not just learning vocabulary.

From *Getting Started in Science: A Blueprint for Elementary School Science Education*, (1989). National Center for Improving Science Education. Andover, MA: The NETWORK.

- High interest: especially at the elementary and middle/junior high levels, the science topics and activities selected are of inherent interest to students.
- Hands-on science: concrete materials are used as a basis for learning at all levels.
- Building on prior understandings: a constructivist approach to learning is fostered, i.e., a minds-on as well as hands-on approach.
- Inquiry teaching: activities are employed which lead to answers to the questions being addressed; laboratory and class activities address important questions.
- Cooperative learning: collaborative learning is employed to promote more intensive and productive study.
- Problem-solving: significant attention is given to higher order thinking skills and finding solutions to problems.
- Accurate knowledge: the science presented is accurate science.

Issues. The issues raised involve questions of what constitutes quality and how to attain it. At the higher grade levels the questions have a different emphasis, due to the more specialized science background of the typical teachers. The following questions stand out.

- What topics should be selected for in-depth study?
- What topics and activities will generate high student interest at the various grade levels?
- How will elementary school teachers be supplied with the necessary hands-on materials on a continuing basis? How will the needed equipment and supplies be provided for science teachers in the middle/junior high and senior high schools?
- How will teachers develop the confidence and ability to teach in an inquiry manner?
- How will teachers be provided the necessary materials and teaching skills for employing cooperative learning in their classes?

- How will a problem-solving focus and attention to higher order thinking skills be developed?
- How will a balance be struck between depth and breadth of study?
- How will teachers learn to avoid excessive abstractness and teach in a constructivist manner?
- How will a problem-solving focus and attention to higher order thinking skills be developed?
- To what extent should the middle/junior high and senior high school curriculum be interdisciplinary, rather than organized around the disciplines?

Appropriateness

Key elements. An appropriate science curriculum gives balanced attention to several dimensions of science content. Students going in a variety of directions after high school require not only a sufficient quantity of science as addressed above, but science that is appropriate to these various destinations. The curriculum content will have an appropriate balance of the following.

- Common themes and concepts from the several science disciplines are presented in an integrated manner that portrays the interdisciplinary and interrelated nature of science.
- Connections are made to other areas of the curriculum, such as writing, mathematics, and reading.
- The science content is connected to the development of attitudes and thinking skills.
- The science content emphasizes major concepts; i.e., the most important aspects of science knowledge are developed.
- Science inquiry is prominent, i.e., the means by which science knowledge is acquired are addressed.

Sidebar 2.3

A FRAMEWORK FOR CURRICULUM

A curricular framework should shape, but not determine, the particulars of what will be taught. The Center's proposed framework consists of organizing scientific concepts, attitudes, and skills. The concepts are intended to give purpose and direction to the design of a K-6 set of experiences that include the life sciences, the physical sciences, and technology, which will enable students eventually to understand the concepts.

Scientific Concepts

1. *Organization (or orderliness).* Science is a human invention, and scientists have made the study of science manageable by organizing and classifying natural phenomenon. . . .

2. *Cause and effect.* Nature usually behaves in predictable ways. Searching for causes and effects and explanations is the major activity of science; effects cannot occur without causes. . . .

3. *Systems.* In science, the concept of a system describes the movement of matter, energy, and information through defined pathways. . . .

4. *Scale.* Scale refers to quantity, both relative and absolute.

5. *Models.* The concept of a model is important to both science and technology. . . .

6. *Change.* The natural world continually changes although some objects or species seem unchanging because of human inability to perceive the rate or scale of change. . . .

From *Getting Started in Science: A Blueprint for Elementary School Science Education* (p. 13-16) by National Center for Improving Science Education, 1989 Andover, MA: The NETWORK.

7. *Structure and function.* There is a relationship between the way organisms and objects look (feel, smell, sound, taste) and the things they do. . . .

8. *Discontinuous and continuous properties (variations).* To understand the difficult concept of organic evolution and the statistical nature of the world, students first need to understand the notion of properties, which in turn involves some understanding of continuous (and discontinuous) variations. . . .

9. *Diversity.* Diversity is the most obvious characteristic of the natural world; even preschoolers know that there are many types of objects and organisms. . . .

- The applications of science to technology, societal issues, personal student needs, and careers are given significant attention.
- Significant attention is given to problem solving and opportunities are provided for students to seek optimum solutions to situations involving complex decisions

Issues. The process of getting these key elements in place potentially raises a number of issues that grow out of varied definitions of appropriateness and questions of how to attain the desired state. These substantial issues include the following.

- What basis will be used for selecting the particular technological topics—and the number of them—that will be included in the science curriculum?
- What basis will be used for selecting the particular student activities on which the program will be built?
- To what degree will science be integrated with language development in elementary schools?
- How will the interdisciplinary aspects of science be developed and connections made to other subjects in the curriculum?

- To what extent should the science curriculum be differentiated for different kinds of students?
- To what extent should the science curriculum include technology and the applications of science to societal issues and the personal needs of students?
- How will the "preparation ethic" be accommodated within an educational context requiring attention to other important goals?
- What applications of science will be included within the content of the curriculum?
- What aspects of the processes of science will be included in the content of the curriculum?
- What topics should be selected for in-depth study?

Conditions and Requirements for Change

Once a desired character for a school science program has been identified and the related issues have been resolved sufficiently, it may appear that significant improvement is close at hand. By itself, however, this step is far from sufficient. The means for getting to the desired state rarely are obvious in any given situation.

While the remainder of this book largely is devoted to this question of how to get there, some broad generalizations are in order here about the overall process of making the desired changes. These key points about attaining desired changes are the product of a cost-effectiveness analysis of the major recommendations for action which have come from various study commissions, recommending groups, and other advocates for change (Anderson, 1990). Sixty nine different recommended actions for science education improvement were analyzed in terms of their probable cost and effectiveness. In addition to specific ratings of cost and effectiveness for each action, the analysis produced some broader generalizations as well.

The key conclusion of this analysis of interventions for improving science education pertained to the systemic nature of the situation. Few single actions intended to produce change gave significant hope; moreover, the many possible interventions are interdependent and interactive. There is real hope, however, for finding some *combination* of interventions which will produce substantial change.

This systemic character of the situation is the key to positive change. It is systemic in numerous ways. The objectives of an educational improvement endeavor—i.e., increases in quantity, quality and appropriateness—cannot be addressed independently. The various levels of education: federal, state, district and school, are highly interactive. Furthermore, an intervention for bringing about improvement in one aspect of the educational enterprise necessarily will affect many others. One generally cannot produce significant change simply by introducing a single intervention.

Second, essentially all interventions have a cost; the idea that an apparently simple action such as increasing graduation requirements in science can be done at no cost is fallacious. The cost of converting a regular classroom to a science laboratory to accommodate increased science enrollments is no small matter. Careful analysis often shows the cost of an intervention to be substantially greater than that assumed by even the seasoned observer.

Third, the cost of developing new curriculum materials generally is relatively low on a per-pupil basis, at least if the materials can be used by a substantial number of people after their development. This cost level is fortunate, given the central role of materials in dealing with the appropriateness dimension described earlier. Significant progress is unlikely without curriculum development, although significant progress on this objective cannot be expected from this action alone; many other related actions will be needed.

Fourth, in addition to actions intended to directly accomplish improvement in the quantity, quality, or appropriateness of science education, attention should be given to facilitating or enabling actions, i.e., supportive actions such as developing local leadership, improving testing programs, and educating the public. None of these enabling actions is likely to be very effective as a *single* action. On the other hand, they may be *necessary* to directly accomplish one of the quantity, quality, or appropriateness objectives. They are needed in an overall systemic effort. Here is where the research on effective schools, implementing educational change, and effective leadership has something to say.

Fifth, many of these enabling actions, such as developing local leadership, are relatively low cost and, when done in concert with other actions as part of a systemic endeavor, quite effective.

Sixth, some important actions potentially take place at several levels: federal, state, local school district, and individual school levels. Even though significant actions at all of these levels are important, the research on effective schools and related studies point to the crucial nature of much that occurs at the level of the individual school. Furthermore, a collection of interventions compatible with the research on effective schools is low cost, compared to many other interventions, and their effectiveness comparatively high.

Taken together, the results of this analysis point to the local level as the place where significant results can be expected. There not only is a desired state for science education that is considerably better than typical practice as found in most schools, but there is reason for a high degree of optimism about the potential of reaching this desired state on the part of local school district and individual school leaders who are informed and take the initiative in implementing change.

References

Aldridge, B.G. (1989). Scope, sequence and sequencing of secondary school science, (Report). Place of presentation: National Science Teachers Association.

American Association for the Advancement of Science. (1989). *Science for All Americans*. Washington, DC: American Association for the Advancement of Science.

Anderson, R.D. (1990). Policy decisions on improving science education: A cost-effectiveness analysis. *Journal of Research in Science Teaching, 27*(6), 553–574.

Bielenberg, Joy. (1993, November) Personal communication.

Harms, Norris C. (1981), *What Research Says to the Science Teacher* (Vol. 3), Washington, DC: National Science Teachers Association.

Linn, Marcia C., *Establishing a Research Base for Science Education: Challenges, Trends, and Recommendations*, Berkeley, CA: Lawrence Hall of Science, 1986.

National Center for Improving Science Education (1989). (*Getting started in science: A Blueprint for Elementary School Science Education*, Andover MA: The NETWORK.

Ogden, William R. (1976). The effect of high school chemistry upon achievement in college chemistry: A summary. *School Science and Mathematics, 76*(2), 122-126.

Pines, A. Leon (1985). Toward a taxonomy of conceptual relations and the implications for the evaluation of cognitive structures. In Leo H. T. West, & A. Leon Pines (Eds.), *Cognitive Structure and Conceptual Change* (p. 194). Orlando: Academic Press.

Shuell, Thomas J., "Knowledge Representation, Cognitive Structure, and School Learning: A Historical Perspective," in West, Leo H.T., and Pines, A. Leon, (Eds.), *Cognitive Structure and Conceptual Change*, Orlando, FL: Academic Press, 1985.

West, Leo H.T. & Pines, A. Leon. (1985). *Cognitive Structure and Conceptual Change*, Orlando: Academic Press.

Wittrock, M.C., "Learning Science by Generating New Conceptions from Old Ideas," in West, Leo H. T., and Pines, A. Leon, (Eds.), *Cognitive Structure and Conceptual Change*, Orlando, FL: Academic Press, 1985.

Woodward, C.A. & R.G. McAuley. (1983). *Canadian Medical Association Journal*, 129.

Yager, R.E. & Krajcik, J.S. (1989). Success of students in a college physics course with and without experiencing a high school course. *Journal of Research in Science Teaching, 26*(7), 599–608.

Yager, R.E., Snider, B., & Krajcik, J. (1988). Relative success in college chemistry for students who experienced a high school course in chemistry and those who had not. *Journal of Research in Science Teaching, 25*(5), 385–394.

3

Establishing the Need to Improve at the Local Level

Why do a Needs Assessment?

The superintendent and science department in Chapter 1 were motivated by a magazine report to begin looking at the local science program. In effect, they had conducted a very quick and intuitive needs assessment. The article "hit home." The description of the national standards was clear enough that they could recognize the difference between what the standards said *should* be happening and what currently was going on in their own school.

Identifying this difference or discrepancy between what currently is happening and what people want to happen (the desired state) is the essence of doing a needs assessment. Finding out what "needs" to be done is the first step in developing an improvement effort. In doing so, several questions must be answered.

Is there a need for improvement?

Who believes that there is a need?

What set of actions will solve the identified need?

In Chapter 2 a number of groups identified the major needs of science education at the national level. In the Chapter 1 scenario a summary of those reports and the emergence of the national standards in the national media motivated the superintendent, the local department chairperson, and a committee of teachers in one of the schools. But it takes more than just these people to solve the problem. There are many actors in the school improvement effort, including the superintendent, school board, school principals, teachers union, classroom teachers, and parents in the community. Rarely are all of the people convinced simultaneously that an improvement effort should occur, especially if there are significant costs involved.

A needs assessment is a way to find out what needs to be fixed and who believes it needs to be fixed. As the old adage states, "If it isn't broken, don't fix it." Often some people think it's broken; others think it isn't. A needs assessment provides a way to identify these differences and to proceed with some action or change in the system; or to decide that nothing needs to be done, the status quo is satisfactory, and it "isn't broken." Maybe the teachers in the science department think that the test scores are as good as they can be, and the principal and counselors think they could be better. A needs assessment identifies how different groups perceive the same issue.

Even if agreement is reached on what the needs are, it is often necessary to define the problem more clearly by breaking it down into smaller parts before it can be solved. As an example, if student achievement in science is poor (less than what is desired) do teachers and administrators think the curriculum should be improved? Or should instructional methods used by teachers be upgraded? Maybe the tests are inadequate. A needs assessment is a way to sort out perceptions about the nature of a problem and how it can be solved.

Starting with Goals

The most important element in all needs assessments is that of goal orientation. It is of little value to identify today's needs without a vision or a goal of where a particular school, community or state wants to be headed (See Chapter 2.). Having a goal helps avoid the focus on just the problems and needs of today.

Any activity, undertaking, or improvement effort needs a goal or set of goals. Often these are unstated or unexamined, but it is essential that a set of goals be established before a needs assessment or improvement project is initiated. There are a number of reasons for setting goals or establishing a vision before undertaking activities. First, the goals ultimately determine the behavior of a group or of an individual teacher in a classroom. As an example, a teacher whose students are expected to remember and follow the information, directions, and definitions presented in a lecture or written format and who rarely uses a laboratory activity—or uses the laboratory only for confirmation activities—has a very different set of goals for his or her students than the teacher who bases student learning upon a set of activities and laboratory experiences in which students are allowed to determine how they will set up the experiment and interpret their own conclusions. The content and concepts presented in the latter class are based upon and anchored in the experiments of the students. Although students may draw their own conclusions, they must be well substantiated by their experiments, the experiments of others, and other sources of information. The questions asked in this class and on the exams have a variety of levels, many of which are well beyond the recall level. The goals of these two individuals give rise to a very different set of personal behaviors and subsequent behaviors on the part of their students. If a group of teachers, administrators, parents, and students are working on an improvement project, the outcome of the project, i.e., the behaviors of all involved, should be result of a well-thought out and clearly stated set of goals, established early in the process.

A second reason for establishing goals is the need to communicate the purpose and need for the improvement effort to individuals and groups beyond the development or planning group. It is often necessary for the planning group to communicate to other teachers in the system, administrators in the central office, the board of education, or members of the public. Without clear goals, this communication will be difficult, if not impossible. Other teachers will want to know what the project is all about. Administrators must decide what support to give the effort. The superintendent and school board must decide whether the project deserves their financial support. Good decisions require having a clear set of goals, well understood and well communicated.

Sidebar 3.1

WHY IS VISION SO IMPORTANT?

A Vision:

- Provides a common reference point for participants with different perspectives in the reform effort
- Can help raise public expectations for both education system and student performance
- Helps gain support for reform by showing people how reform will look and how their roles in the education system and relationships with students and one another will be different
- Builds persistence. People committed to a common vision do not give easily in the face of opposition.
- Provides indicators by which to measure progress
- Helps motivate, inspire and call people to action
- Is a tool that allows a community to control its own destiny

A third reason for having a well developed set of goals is to avoid the confusion of means and ends. The goal of improving student achievement probably has a number of alternate means of attainment. Will a new curriculum help? Will more computers and better software improve student achievement? In both cases, only if they are aligned with the outcomes measured on the achievement tests. The means may be important and have a variety of positive benefits, but are they a way of getting to the end; i.e., the stated goals? Often the means—such as a new set of computers—becomes an end in itself because of the excitement and innovative nature of the activity. But we should not confuse such means with achievement of the stated goal.

With the importance of goals in mind, let's examine their source. The elements of a desired state of science education described in Chapter 2 are essentially goals. Goals do not have any particular form of statement, nor are they precise statements that can be achieved on short-term bases. Short-term matters are better described as objectives. Goals are targets that will take some time and effort to reach.

An improvement effort is a set of activities designed to reach unmet goals. The goals may be old ones that an evaluation or needs assessment has indicated are not yet met. On the other hand an improvement effort may be focused on new goals that have not yet been clearly identified or accepted by the local school or district. Examine some of the elements or goal statements in Chapter Two. They may not be new to the reader but if they have not been written and accepted by the members of the science department or school district, they should be considered new goals and ready-made targets for an improvement activity.

How can goals be developed by a local school or district? The process is probably best accomplished by a small committee that has opportunities to interact with—and receive input and reaction from—all other members of the system.

A typical way to proceed would be to convene a committee of eight to twelve members who represent the various constituencies of the system. Is this a K-12 effort? Should all subject matter areas be represented? What about principals? Parents? Students? Are the members of the committee simply a voice of these groups, or should they have a means of soliciting input from each of these constituencies and in turn carrying the work of the committee back to them? In other words, does the biology teacher on the committee have an opportunity to carry information to all biology teachers in the district and accept instructions from them, or is he or she simply a spokesperson for that group?

Once the committee or task force has been convened, provide the committee with goals that have been developed by other school districts or national organizations. Figure 3.1 contains five sets of goals from: (1) the National Science Education Standards, (2) Project 2061, (3) the National Science Teachers Association, (4) Project Synthesis, a research project supported by the National Science Foundation, and (5) an individual school district. What does the committee find that these lists have in common? Often similar goals appear on more than one list. What did they leave out? When were they developed? Goals change over time, so it is important to note the date of each document.

FIGURE 3.1
Sample Goal Statements

National Science Education Goals(1)

The ultimate goal of the National Standards for Science Education is a scientifically literate citizen that enables a democratic government, a healthy economy, and viable scientific and technological communities. Being scientifically literate empowers individuals

to engage intelligently in public debates around matters with scientific and technological aspects,

to increase their economic productivity

to use scientific principles and processes in making personal decisions, and

to respect and enjoy the natural environment.

Science For All Americans

AAAS Project 2061(2)

Scientific literacy—which embraces literacy in science, mathematics, and technology—has emerged as a central goal of education. It includes being familiar with the natural world and respecting its unity; being aware of some of the important ways in which mathematics, technology, and the sciences depend upon one another; understanding some of the key concepts and principles of science; having a capacity for scientific ways of thinking; knowing that science, mathematics and technology are human enterprises, and knowing what that implies about their strengths and limitations; and being able to use scientific knowledge and ways of thinking for personal and social purposes.

Science — Technology — Society:
Science Education for the 1980's
NSTA Position Statement (3)

1. Develop scientific and technological process and inquiry skills
2. Provide scientific and technological knowledge
3. Use the skills and knowledge of science and technology as they apply to personal and social decisions
4. Enhance the development of attitudes, values, and appreciation of science and technology
5. Study the interactions among science-technology-society in the context of science-related societal issues

---

Project Synthesis (4)

Goal Cluster I: Personal Needs
Prepare individuals to utilize science for improving their own lives and for coping with an increasingly technological world.

Goal Cluster II: Societal Needs
Produce informed citizens prepared to deal responsibly with science-related societal issues.

Goal Cluster III: Academic Preparation
Allow students who are likely to pursue science academically as well as professionally to acquire the academic knowledge appropriate for their needs.

Goal Cluster IV: Career Education/Awareness
Give students an awareness of the nature and scope of a wide variety of science and technology-related careers.

Science Program Goals (5)

Jefferson County (CO) Public Schools

The elementary and secondary science program in the Jefferson County Public Schools will:

1. Develop scientific and technological literacy for all students.

2. Provide an academic course of study that will:
   a. Encourage students who are likely to pursue science-related professions to acquire the knowledge appropriate for these needs (professional preparation).
   b. Prepare individuals to utilize science for improving their own lives and coping with an increasingly technological world (personal needs).
   c. Produce informed citizens prepared to deal responsibly with science-related social issues (societal issues).
   d. Give all students an awareness of the nature and scope of a wide variety of science and technologically related careers open to people of varying aptitudes and interests (career awareness).

3. Develop in students the decision making skills which enable them to apply scientific and technological knowledge in solving personal societal problems.

4. Promote the student's self image and a positive attitude toward science.

5. Provide science instruction that matches students' mental, physical, social and emotional growth and is consistent with their future plans.

Notes

1. National Research Council (1992) *National Science Education Standards: A Sampler*, Washington, DC: National Research Council.

2. Rutherford, J.R. & Ahlgren, A. (1993) *Science for All Americans,* New York: Oxford University Press.

3. NSTA, (1982) *An NSTA Position Statement: Science-Technology-Society: Science Education for the 1980's,* Washington, DC: NSTA.

4. *What Research Says To The Science Teacher. Vol. 3,* Washington, DC: NSTA. 1981 (Revised 1985)

5. H. Pratt, Personal correspondence.

After the committee has read through and reviewed the sample goals, find a facilitator who can lead the group in brainstorming what they believe their goals should be. This list will obviously be influenced by the published lists, so you may choose to do this activity first before reviewing the already documented goals.

Regardless of the order in which the first two activities occur, merge the two lists and reduce it to the smallest number of statements possible without eliminating important ideas. At this point the committee needs to decide if it wants to pare the list back any further, or if it is time to go out to their various audiences in the district and solicit their input.

Although it may not be possible, we highly recommend that this be done in small group meetings, where a detailed discussion of each goal can occur and suggestions for rewording and questions of clarification be addressed. At the end of such a session, some form of written survey or feedback should be collected. Everyone should have an opportunity to indicate the degree to which they support each goal. This will allow the goals committee an opportunity to collect data on the degree of acceptance of each goal.

Once these data have been summarized for all audiences, the committee should decide if any goals need to be rewritten or discarded. Once this decision is made the results should be published and disseminated to everyone involved.

Official approval of the goals by the Board of Education is strongly recommended. This not only provides an official sanction or approval of these goals; it is a golden opportunity to communicate and explain the goals to the key decision makers in the district. Later, when recommendations for funding or approval of projects are brought to the board, it can be pointed out how these plans have been developed as a means of meeting the goals previously approved by the Board of Education.

Although we do not believe there is one right way or form in which goals should be written, here are a few hints for improving the quality and clarity of the goals. (See the National Education Goals as one example in Chapter 1.)

1. Keep the list short and the statement of any individual goal at a reasonable length.
2. Describe some kind of outcome, product or behavior that is reasonably clear. Although not objectives, they should create an image in the minds of the reader of what would exist, or how students would act, when the goals are met.
3. Avoid confusing means and ends. As discussed earlier, the process of meeting a goal often is confused with the goal itself. Providing better science facilities is not a goal that usually will be found in many lists, but it is an appropriate means of improving the quality, quantity or the style of teaching and learning that occurs in a school.

The critics or resisters of such a goal development process should not be allowed to stall the process by arguing that the goals developed several years ago have not yet been met so why should we develop a new set of goals. Developing a new set of goals is not like advancing the target further down range or moving the carrot in front of the racing rabbit ahead of him at a faster rate. Instead, we are asking if we have new targets located in new territory altogether different than the old ones. We live in a rapidly changing world that should be causing us to rethink our goals on a regular basis. Not all of yesterday's goals are appropriate in a rapidly changing culture.

Starting with Strengths or Weaknesses

One perspective is to start by doing a "strength assessment" by asking: "What is working well?" "Why is it working?" "What do

Sidebar 3.2

GOALS OF SCIENCE EDUCATION HISTORICAL PERSPECTIVE

1900 – 1960	Know principles of science. Test principles in laboratory.
1960 – 1980	Know principles of science. Test principles in laboratory Experience the process of gaining knowledge.
1980 – 1990	Know principles of science. Test principles in laboratory. Experience the process of gaining knowledge. Understand the relationship of science and society.
1990 –	Construct an understanding of science principles Develop the skills, habits of mind, and understanding necessary to conduct an inquiry. Use science principle to resolve personal and societal problems.

we need to do differently (i.e., more of or less of) to be even more effective?" "Which of these actions is our highest priority?" This provides a real morale boost for the staff and community members involved and keeps the outcome from constantly focusing on how bad things are. After a committee at a school identifies the many things it does best, it can then begin to work down the list in a very positive way and identify those things that still need some improvement.

A more traditional way to proceed with the needs assessment is to have a group of people take stock of the current status of a program and define the places where improvement is needed. In other words, where is it broken, where are the problems?

Adapted from Brinckeroff, Richard, Personal Communication.

These procedures focus on the basic idea that a "need" is the difference between the desired state and the actual state of a program or school. In the first perspective, the desired state was projected into the future so that the system tends to move toward a more future oriented goal. In this case the need for goals and their use comes early in the process.

In the first perspective, the group starts with features where the discrepancy is small and therefore the need is very limited and goes from there to topics where the discrepancy is greater. Goals are implied in each of the items but are not stated explicitly. In this case goals should be carefully identified in the process of developing the improvement targets.

Collecting Data

The actual state of the science program is determined by assessing the current state of affairs within the district. But it is guided by the newly developed goals or a description of the desired state that has been developed for the improvement project. Since our goal or purpose is to move toward an improvement, it may not be necessary to collect data on all dimensions of the current status. As an example, if you are no longer interested in a goal of students achieving at a certain level on a standardized test of science knowledge, reporting student scores for the last five years on this test would be unnecessary data. Goals and vision will guide you in deciding what type of data to collect from which audiences. The data collection process needs this direction; otherwise, too much data will be collected from too many people without focus or purpose, leading only to extra work and possible confusion.

The types of data that can be used to provide information about the actual state of your science education program range from standardized test data to informal reports and may include even letters from parents to the school board. There are at least two purposes for collecting the data: 1) to test the assumptions, goals, and values expressed by the description of the desired state of your science education program and 2) to determine where your existing program is with respect to the desired state. With this in mind, all

Sidebar 3.3

Creative Tension
A Leadership Principle Similar to a Needs Assessment

Peter Senge, author of *The Fifth Discipline: The Art and Practice of the Learning Organization* describes how leadership in a learning organization starts with principle of creative tension.

> Creative tension comes from seeing clearly where we want to be, our 'vision,' and telling the truth about where we are, our 'current reality.' The gap between the two generates a natural tension. Creative tension can be resolved in two basic ways: by raising current reality toward the vision, or by lowering the vision toward current reality....
>
> Without vision there is no creative tension. Creative tension cannot be generated from current reality alone. All the analysis in the world will never generate a vision. Many who are otherwise qualified to lead fail to do so because they try to substitute analysis for vision.... What they never grasp is that the natural energy for changing reality comes from holding a picture of what might be that is more important to people than what is.
>
> But creative tension cannot be generated from vision alone; it demands an accurate picture of current reality as well. Vision without understanding of current reality will more likely foster cynicism than creativity. The principle of creative tension teaches that an accurate picture of current reality is just as important as a compelling picture of a desired future.

From P.M. Senge (1990) The leader's work: Building learning organizations, *Sloan Management Review*, 32(1).

the data, questionnaires, test scores, and informal responses should be focused on the desired state.

Some of the sources or types of data that you could use are as follows.

International, National, or State Reports

These reports may or may not apply to your local school situation.

Call your state department of education and ask for recent reports on the quality of science education in your state. Have they developed goals or long range plans for upgrading science education in the future? Two recent studies are typical of the data that are often available from the National Science Foundation, National Assessment of Educational Progress (NAEP) and other groups. These are: *Indicators of Science and Mathematics Education-1992* (NSF,1992) and Learning *Science: International Assessment of Educational Progress* (Lapointe, Askew & Mead, 1992).

Virtually all states and school districts have some form of testing program from which data can be extracted. Keep in mind the similarities and differences between the local school population and that of the population for which the data were reported. Also, it is important to recognize that most tests can measure only a small portion of the goals and outcomes your desired program probably includes. If one of your desired outcomes is higher level thinking skills, and the tests only measure the recall of information at a lower level, don't be misled into thinking that you have adequate data.

Sort the data very carefully and look at all parts of the student achievement results. As an example, the recent National Assessment of Educational Progress indicated that the science proficiency of the nation's students declined for several years from 1970 until 1982 and has been increasing slowly since then.(See Figure 3.2) Are comparable data available in your district or state?

Check the data for information on various subpopulations. Do girls do as well as boys? What about minority groups? Can we tell the effect of students who have only recently moved into the school district versus those who have been the district for many years? In other words, guard against simply looking at a single mean score for a given population of students in a given subject area.

Sidebar 3.4

Science Achievement in Seventeen Countries* Rank order of countries for achievement at each level

	10 yr. olds Grade 4/5	14 yr. olds Grade 8/9	Grade 12/13 Science Students Biology	Chemistry	Physics
Australia	9	10	9	6	8
Canada (Eng)	6	4	11	12	11
England	12	11	2	2	2
Finland	3	5	7	13	12
Hong Kong	13	16	5	1	1
Hungary	5	1	3	5	3
Italy	7	11	12	10	13
Japan	1	2	10	4	4
Korea	1	7	–	–	–
Netherlands	–	3	–	–	–
Norway	10	9	6	8	6
Philippines	15	17	–	–	–
Poland	11	7	4	7	7
Singapore	13	14	1	3	5
Sweden	4	6	8	9	10
Thailand	–	14	–	–	–
USA	8	14	13	11	9
total # of countries	15	17	13	13	13

* Pergamon Press, 1984

Sidebar 3.5

The most recent science information is available from the National Assessment of Educational Progress. For an application form and information on how to obtain NAEP items and data, contact:

Bob Clemens, Coordinator of NAEP Items

U.S. Department of Education

National Center for Education Statistics

Education Assessment Division

Room 308

555 New Jersey Avenue, N.W.

Washington, D.C. 20208-5653

Phone: 202-219-1729

Graduate Follow-up Studies

Has your school district or state done a follow-up study of graduates 1 year, 5 years, or 10 years after they completed high school? There may be some very valuable data imbedded in this report, or you may be able to add a few questions to the next survey, focused on some dimension from your desired state on which you would like some information from recent graduates.

PTA or Advisory Committee Minutes and Reports

Do any of your schools have advisory committees or PTA's that have spent time studying the quality of education in their school? The reports may be of great value to you.

Figure 3.2

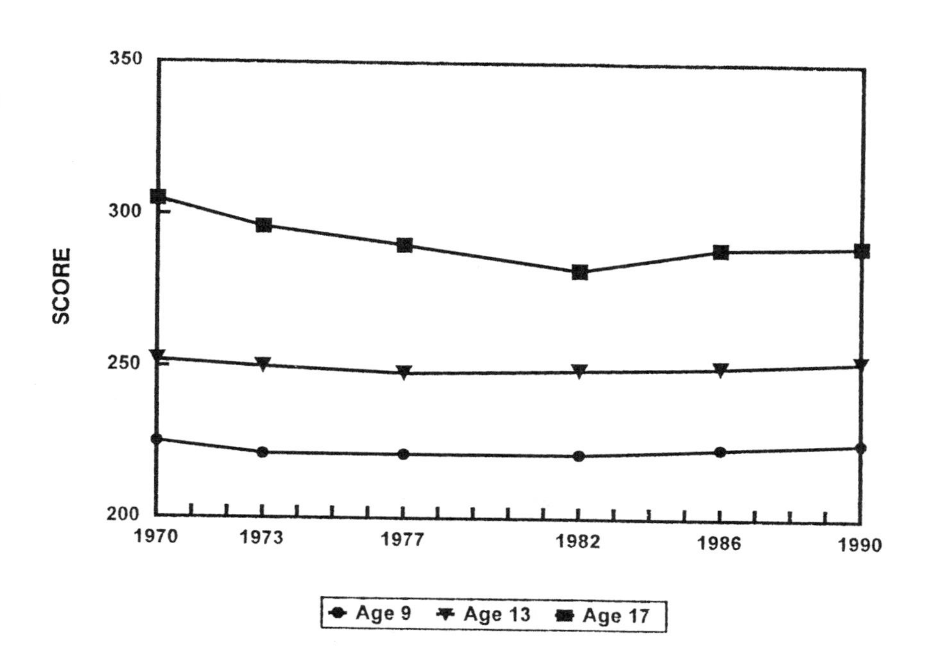

It will undoubtedly be necessary to collect some original data based upon instruments and procedures that you will need to create. Consider the following possibilities.

Student Attitude Surveys

Have any of the schools or student councils conducted a student attitude survey with questions related to science? Are such data available at the state level? The NAEP data include extensive student attitude information that may be very applicable to your district.

From National Center for Education Statistics, *Trends in Academic Progress*. Report No. 21-T-01 (Washington, DC: U.S. Department of Education, 1991).

Sidebar 3.6

The following is a somewhat simplified example of how three questionnaire items lead to an improvement goal.

Indicate your priority for the content in middle school science courses.

a.	Content necessary to deal with personal needs such as use of drugs, alcohol, personal appearance, nutrition,and sexual behavior.	1, 2, 3, 4, 5 (1 = highest)
b.	Content necessary to prepare for the next level science course.	1, 2, 3, 4, 5 (1 = highest)
c.	Content that satisfies the curiosity of students by helping them understand the natural world.	1, 2, 3, 4, 5 (1 = highest)

National and Local Opinion Polls

The Gallup Poll annually collects and publishes data in *Phi Delta Kappan* (Elam, Rose & Gallup, 1993) on the opinions and attitudes of parents toward a number of major issues in education. Some of these may be applicable to your needs. Similar data may be available within your state or local school district.

Questionnaires

This is the obvious source of data for a needs assessment, but it needs to be carefully designed. Poorly stated questions will produce responses which only increase the problem, not clarify the needs. Questions should be unambiguous, brief and relate to a single idea. Questions should be directly focused on one of the ideas from the desired state. For example rate the importance of students being able to solve problems on a scale from very unimportant to very important. Consider using questions which require respondents to rank order several outcomes, or to give each of them a weight according to how much instructional time should be spent on each outcome. It is too easy to respond positively to all the ideas on a questionnaire. The real test is to find out which of the ideas are most important. If instructional time and resources are limited, as they always are, then test the respondents to find out how they would allocate time and money for each of the desired outcomes. See Figure 3.3 for an example of this type of item.

Figure 3.3

What would be your recommended distribution of time and resources to the four goals or areas of emphasis? Each column in the table below represents a distribution of time and resources. Please study the table and then answer the three questions that follow.

	Percent of *Time and Resources*				
	a	b	c	d	e
Personal Needs	5	15	55	15	30
Societal Issues	5	15	15	55	30
Academic Preparation	85	55	15	15	30
Career Ed. Awareness	5	15	15	15	10
	100%	100%	100%	100%	100%

From the information in Table 1 above, select the distribution which comes closest to what you believe should be the recommended distribution of time and resources.

Mark the *best* distribution for:					
Elementary Science	a	b	c	d	e
Junior High Science	a	b	c	d	e
Senior High Science	a	b	c	d	e

The audiences for such questionnaires usually include teachers, administrators, parents and students. Keep the data separate for these groups; there is little rationale in combining or averaging the response from 300 students with those of 30 parents and ten teachers.

Sidebar 3.7

One of the best questionnaires for assessing your science program that we know of is the *Guidelines for Self Assessment* produced by the National Science Teacher's Association, 1840 Wilson Boulevard, Arlington, VA 22201-3000. It comes in three versions: elementary, middle/junior high, and high school programs.

The respondents mark each item in the survey on a matrix, indicating both the desirability and achievement of given statement. Such a process, which can be hand or computer scored, immediately indicates the discrepancy between the current or actual state of things compared to the desired state.

If you have not already done so, the questionnaire is a good opportunity to check out the assumptions, goals and values of the desired state since they probably were developed by a small, select committee. You are not compelled to follow the opinions of all teachers, students or citizens, but it is very important to know where the members of each of these audiences stand with respect to your future outcomes. This information will provide much important insight on the amount of time, effort, and resources necessary to bring about the improvements or changes needed.

External Audit Team

For an objective point of view many school districts are beginning to use an external audit team (Baum & Brady, 1986) to review school or school district programs, such as science. The North Central Association of Secondary Schools has undertaken an experimental review process to evaluate a single discipline such as K-12 science for an entire school district, in addition to their traditional review of individual schools. In other cases, districts have established their own processes.

Although the audit team can be made up in a variety of ways, one pattern is a team composed of local citizens within the district, teachers and science educators from outside the district, scientists, doctors, engineers, etc. who are employed or reside in the district, and a chairperson of the group who is a recognized expert in science education.

Once the data are collected and summarized, the original needs assessment committee should carefully and thoroughly review the data and develop a series of goals or visions for the improvement effort. These goals should portray a description of what success would look like when the improvement effort has been completed. How will you know that the desired state has been met? What will you be able to see, observe and measure when you reach that point? In the next chapter we will discuss the process of moving from goals to the development of activities that support and meet these goals.

References

Baum, E. & Brady, K.P. (1986). A working model for external audit of instructional programs. *Educational Leadership, 43*(5),

Elam, S.M., Rose, L.C., Gallup, A.M. (1993) The 25th annual phi delta kappa/ gallup poll of the public's attitude toward public schools. *The Phi Delta Kappan, 75*(2),

Lapointe, A.E., Askew, J.M., & Mead, N.A. (1992) *Learning Science: The International Assessment of Education Progress*, Princeton, N.J., Educational Testing Service.

Division of Research, Evaluation and Dissemination, Directorate for Education and Human Resources. (1993) *Indicators of Science and Mathematics Education 1992.* Larry E. Suter (Ed.) Washington D.C: National Science Foundation.

Senge, P.M. (1990) The leader's work: building learning organizations. *Sloan Management Review, 32*(1) 1990

4

Acquiring Teaching Materials

Reaching the new state of science education you have set as the goal of your school or school district is a multi-faceted and long-term endeavor. Two facets are particularly prominent—getting the appropriate teaching materials in place and helping teachers to develop approaches to instruction that are consistent with the new goals and teaching materials. Curriculum materials and instructional practices are intimately intertwined; it is difficult to address one without the other. For convenience, however, we will address teaching materials in this chapter—with occasional references to teaching approaches—and then move directly to means of instructional improvement in Chapter 5. These two facets—along with all the others—cannot be attacked independently in actual practice; we are addressing a systemic situation that needs systemic approaches to change.

Putting new teaching materials into practice appears to be the most common beginning point for attempting science education change. This situation is not surprising given the central role that textbooks and related materials play in science teaching and the heavy dependence teachers place on them as the guide for what will be taught. While one may argue about the desirability of this prominent role for textbooks, this role is reality.

In examining this situation, some persons enter into a "chicken or egg" discussion. Is it that the available textbooks determine what teachers will teach, or is it that teachers choose those textbooks that contain what they want to teach? Are teachers restricted in what they teach by a limited range of materials, or are publishers only able to sell a limited range of materials because there are too few of buyers

Sidebar 4.1

THE ROLE OF TEXTBOOKS

At least since the days of the McGuffey reader, textbooks have occupied a central role in the educational process. Research shows that textbooks clearly have this central role today in science teaching. In their *Case Studies in Science Education*, Stake and Easley (1978) found a heavy reliance upon textbooks in science classes as the determiner of what is taught,

> Behind nearly every teacher-learner transaction reported in the case studies lay an instructional product waiting to play a dual role as medium and message. They commanded teachers' and learners' attention. In a way they largely dictated curriculum. Curriculum did not venture beyond the boundaries set by the instructional materials.

for any but the most conventional materials? Whatever the outcome of that debate, it is fairly clear that (1) the range of difference among textbooks, in terms of both teaching approaches employed and the aspects of science content taught is fairly limited and (2) these materials are found in the vast majority of American classrooms.

Persons with responsibility for administering and improving district science programs face the question of what teaching materials they should select from among those commercially available—there is, of course, some variation—or alternatively developing their own teaching materials if they are not satisfied with what they see in the market place. These questions comprise the focus of this chapter. In the next chapter, attention will be turned to more direct means of changing instructional practices to coordinate with the new materials and other reflections of a new set of goals.

Textbooks and teaching materials have been used almost interchangeably in this discussion thus far; some sharper distinctions in terminology are needed. Among the vast array of teaching materials that could be considered, three in particular are singled out for attention here: textbooks, curriculum guides, and laboratory materials. Textbooks are the most prominent of these teaching materials in science classes. Curriculum guides, produced by the local school district, often exist but clearly do not have the influence in classrooms that textbooks do, particularly at the secondary level. Laboratory materials—mainly the wide variety of equipment involved, and student laboratory guides—also play a very prominent role in many secondary school science classrooms.

As in the case of secondary schools, these three categories of teaching materials also play important roles in elementary school classes as well, but their relative importance is somewhat different. Textbooks play a central role in the vast majority of—but not all—elementary school classrooms. Curriculum guides seem to have more influence with elementary school teachers than with secondary school teachers, but in most situations the textbooks are most influential. In contrast to secondary schools, laboratory teaching materials play a much lesser role in the typical elementary classroom. There are some important exceptions to this generalization about elementary schools, however, in that a small minority of elementary school teachers are using science programs that are not built upon textbooks, but are centered around an abundance of hands-on student materials.

What teaching materials can be used and how can they be acquired? As a prelude to these specific questions attention must be given to (1) the role of teaching materials in the educational setting, at the elementary and secondary levels, and (2) the nature of the curriculum content.

The Role of Teaching Materials

The role of teaching materials at the elementary school and secondary school levels is different enough that the two will be addressed separately here. In each case attention will be directed to the variety of roles that various materials can play and some of the bases for choosing among them.

Approaches for Using Materials to Teach Elementary Science

Read and talk about science. The most common teaching approach is based on one of the standard textbook series available from several commercial publishers. Most such series include a systematic coverage of a wide variety of science topics in a spiral manner, although organization and grade placement varies considerably from one series to another. Along with the narrative explanation of the various science concepts, one typically finds descriptions of activities that could be conducted with rather simple equipment by either teacher or students. By far the most frequent practice in schools, however, is for teachers to simply read the description with the students and talk about it.

The "read and talk about science" approach under consideration here is very prevalent in American elementary schools because it is easy to do, and is the least threatening to the elementary school teacher who is not overly excited about science. It receives further encouragement because it does not require the school to purchase quantities of hands-on materials and equipment for students.

It is clear, however that this approach to teaching science is far from ideal. Research shows clearly that approaches to science teaching which involve the use of manipulative materials yield greater student gains on measures of science achievement. (Shymansky, Kyle, & Alport, 1983 and Bredderman, 1984). Not surprisingly, the evidence also indicates that students learn more from the hands-on approach in terms of understanding the nature of science, skill in manipulating materials, attitudinal changes, and other educational objectives. The evidence indicates that this most prevalent of approaches to teaching science definitely is not the desired state of science education discussed in the previous chapters.

Textbook plus equipment. A modification of the "read and talk about science" approach is to supplement it with teacher demonstrations of activities found in the textbook, or student replication of the activities, using classroom quantities of the necessary equipment and materials. In most cases the necessary equipment is available for purchase from the publisher of the textbook series or a supplier with which they have made arrangements. This approach has the potential of introducing the missing hands-on element into the standard approach, but often the activities shown in the text-books are not the best for student class activities. The development of such textbook programs generally is focused upon the written materials; the hands-on activities found in the text were chosen to illustrate the written narrative rather than to serve as the foundation of the program—a fundamental and usually disastrous flaw. We do not recommend this approach either.

Commercial hands-on programs. In addition to the standard textbook-based programs from commercial publishers, science programs of the hands-on variety are available, having classroom quantities of student manipulative materials. Originating with the NSF- funded curriculum projects of a quarter century ago, e.g., *Science a Process Approach* (SAPA), *Elementary Science Study (ESS), and Science Curriculum Improvement Study* (SCIS)—some but not all are still available—programs of this type were designed *without* reading materials and are built upon activities specifically designed for student learning by manipulation of hands-on materials. The use of these materials provides the basis for student thinking and discussion, without the use of textbooks for the students to read. The advantages of such programs are greater student learning and interest in science, as demonstrated by the research cited above.

Hands-on programs recently received a big boost with the release of a new generation of elementary science programs developed with NSF funding. Available from commercial publishers, they provide a number of alternatives and give schools a high probability of finding one or more that fit their vision of a quality science program. Although some, but not all, of these programs have student reading materials, they are of quite a different character than conventional elementary school science textbooks. These reading materials are more literary and story-like, without a focus on scientific facts and vocabulary. Schools seeking teaching materials that will provide a solid foundation for an elementary school science program in keeping with the new National Science Education Standards would do well to consider one of these programs. (See Sidebar 4.2)

The disadvantages of programs with many hands-on manipulative materials are illustrated by the modest degree of success that many school districts have had in introducing the previous generation of such programs *and sustaining them* over a period of time. The

Sidebar 4.2

NEW ELEMENTARY SCHOOL SCIENCE PROGRAMS

In the last several years the National Science Foundation has supported the development of a new generation of hands-on elementary science programs. The materials for these new programs (and one program—SCIS—that has continued from the first round of funding) are now available from a variety of publishing companies.

Program and Developer	*Publisher*
Full Option Science System (FOSS) Lawrence Hall of Science University of California Berkeley, CA 94720	Encyclopedia Britannica Educational Corporation
Insights: An Inquiry-Based Elementary School Curriculum Education Development Center 55 Chapel St. Newton, MA 02160	Optical Data Corp.
Kids Net TERC 2067 Massachusetts Ave. Cambridge, MA 02140	National Geographic
Life Lab Science Program 1156 High St. Santa Cruz, CA 95064	Videodiscovery Inc.
Science Curriculum Improvement Study (SCIS) Lawrence Hall of Science University of California Berkeley, CA 94720	Delta Education
Science for Life and Living BSCS 5415 Mark Dabling Blvd. Colorado Springs, CO 80918-3842	Kendall/Hunt Publishing Co.
Science and Technology for Children (STC) National Science Resources Center Arts and Industries Building, Room 1201 Smithsonian Institution Washington, DC 20560	Carolina Biological Supply Co.

support required in the form of equipment (including resupplying expendable materials) and in-service education of teachers was not insignificant. The operation of such programs by some school districts over a period of many years, however, is evidence that it can be done. Most leaders in science education would claim that the greater student outcomes attained are worth the additional commitment and resource allocation required. Furthermore, experience with the earlier generation of such programs has provided a basis for the developers of the new generation of programs to ease these associated difficulties.

Local curriculum guides. A local school district may choose to develop its own curriculum guide for science and use it as the basis for their science program. In some cases the curriculum guide is intended to stand alone, while in other cases it is built on the expectation of student use of reading material in one or more textbooks. They also may vary in terms of the use of hands-on materials by students; some guides are developed for such usage, while others may assume little or even no student use of such materials.

If the locally developed guide is based on the use of appropriate hands-on materials *and* a system of supplying these materials to teachers in a convenient and timely manner is maintained, this approach is another viable way of establishing an elementary school program that is consistent with the National Science Education Standards.

Choosing the Type of Elementary Science Program for your District

The basic choices faced by the leaders responsible for the elementary science program in a school district are reflected in the alternative program approaches described above. Most science education leaders are firmly convinced that the hands-on approaches are superior; the empirical research is firmly in support of this position. If one wishes to have a science program of the "desired" form discussed earlier, a move in this direction will be necessary. The greater implementation challenges of this approach to elementary school science are addressed more fully in later sections of this chapter devoted to purchasing and developing materials.

Approaches for Using Materials for Secondary Science Programs

In secondary science programs, the two most prevalent forms of classroom activities are: (1) teacher presentations and (2) laboratory instruction.

Teacher presentations. By far the most common approach to teaching science in the secondary schools is one in which the teacher does the majority of talking. This talk is directed toward:

1. Conveying science knowledge,
2. Directing class recitations in which students respond to quite specific and direct questions, or
3. Directing the use of textbook exercises or worksheets.

This approach is typically very closely tied to the use of a textbook (Stake and Easley, 1978), a textbook which describes the same science content being explained by the teacher and held out to students as the goal of instruction.

This approach, with only minor variations, has been shown by research (Goodlad, 1983) to be the most prevalent pattern of instruction in U.S. secondary school science classes, probably because it is the established pattern the vast majority of people have experienced in the past. The disadvantages are less student interest in science, somewhat lower achievement, and a restricted set of educational objectives. It is not the "desired state" of science education.

Laboratory instruction. Most secondary school science classes involve the use of scientific equipment for demonstrations and laboratory activity purposes. Two alternatives are presented here because of their substantially different character and the fact that the two together cover most of the variations.

One approach, whether a demonstration or laboratory activity, involves the use of equipment in a manner that enables students to see in a concrete way what they are also learning at the verbal level, and thus verify and strengthen their learning of cognitive material. A substantially different approach to the use of such equipment involves student attempts to develop generalizations or student testing of hypotheses for which they do not yet know the answers. Rather than using these materials simply for verification, they involve an inquiry or discovery approach to science learning.

Some laboratory manuals and guidebooks are said to be based on this inquiry approach; actually, the variation among textbooks and laboratory manuals in this regard is not great. The differences found from one classroom to another (research indicates that these differences are not large) are due more to the approach taken by the secondary school science teacher than by the materials at hand. As a result, there is reason to give attention to educating teachers to use different approaches—together with or independent of introducing new materials—a matter which will be addressed in a later chapter.

Content of the Secondary Curriculum

There is another fundamental question, "What should be the content of the curriculum?" Choices must be made about the emphasis given to such dimensions as (1) science concepts and principles. (2) the interconnections among these concepts and principles in the various science disciplines, as well as between these science concepts and those in other fields such as mathematics and engineering; (3) the methodology by which scientists acquire this information; and (4) the applications of this knowledge in such contexts as technological advancement, the resolution of science related societal issues, and the personal needs of individual students.

Assuming that goals have been set and your "desired state" determined, the next step is to examine some of the curriculum content alternatives available in various teaching materials.

The first example is the standard type of textbook used in the typical secondary school science course. Typical books contain an overview of the major concepts and facts identified for inclusion in the curriculum at that level. Discussion of means by which scientists acquire information in a given field typically is minimal at best. Similarly, little attention is given to the applications of science knowledge. One analysis of textbooks showed that most books devoted less than 5% of the space to the total range of applications of science knowledge of all sorts. Science textbooks generally are developed with the expectation that science knowledge, largely independent of its source or applications, is the curriculum content. (Harms, et. al, 1981)

Another example, gives much more attention to the applications of science knowledge. It is well portrayed by the relatively new chemistry course, *ChemCom* (derived from the more descriptive full course title: *Chemistry in the Community*), developed by the American Chemical Society with financial support from the National Science Foundation. It was hoped that the 40% of high school students then (1985) taking chemistry (largely college bound students) could be expanded considerably by providing an alternative chemistry course, giving more attention to the application of chemistry knowledge.

The course is organized around eight modules or units which have as their focal point some *application* of chemistry knowledge, such as water, petroleum for burning and building various products, health and nutrition, or the chemical industry. Although these topical modules provide the structure of the course, they have a carefully developed sequencing of chemistry knowledge which provides a systematic development of knowledge in a manner similar to what happens in a standard chemistry course. The major differences are that (1) applications form the key organizational structure of the course and (2) more time in this course is devoted to the applications of knowledge than in the conventional course.

In terms of student enrollment, it is clear that *ChemCom* has been a success, and a similar path is being followed in other fields. An analogous course is under development in physics, as well as biology. As in the case of *ChemCom*, these new courses are being funded by the NSF. Published versions of these two new programs are expected in 1995 and 1996. (See Sidebar 4.3.)

A third type of science course, focuses on science-related issues, for example, environmental issues. While science knowledge *per se* is taught in this context, no attempt is made to systematically develop the knowledge of some particular field of science. The focal point is the issues themselves and time is devoted to the teaching of science knowledge only to the extent needed for addressing the particular issues.

These three examples, illustrate the range of types of science knowledge that may be portrayed in a particular set of teaching materials. The examples are from secondary school science, but similar differences can be found in elementary school materials as

Sidebar 4.3

A NEW GENERATION OF APPLICATION-ORIENTED SENIOR HIGH SCHOOL SCIENCE PROGRAMS

A number of new senior high school science courses with an emphasis on application of the discipline have been published or are in development with publication expected in the near future.

Program and Developer	*Publisher*
Active Physics American Association of Physics Teachers and American Institute of Physics 60 Stormytown Rd. Ossining, NY 10562	Developer
BioCom Department of Science Education and Biology Clemson University Clemson, SC 29634	International Thompson Publishers
Biology: A Human Approach BSCS 5415 Mark Dabling Blvd. Colorado Springs, CO 80918-3842	Kendall/Hunt Publishing Co.
ChemCom American Chemical Society 1155 16th St., N.W. Washington, DC 20036	Kendall\Hunt Publishing Co.
Insights in Biology Education Development Center 55 Chapel St. Newton, MA 02160	Developer

well. Differences among elementary school materials may tend to be more a matter of the teaching approaches employed; the science content, however, also varies along some of the dimensions described in these three examples. Secondary school science materials, on the other hand, appear to have greater variation in the type of content presented. In contrast to elementary school materials, these variations among content do not seem to be accompanied by as much variation in teaching approaches.

Purchase, Adapt or Develop?

It is clear that the school district attempting to acquire new teaching materials faces many issues in the process. *It deserves far more attention than simply appointing a committee of teachers to select the best book available.* One needs to decide what type of teaching materials are needed and then evaluate the existing available materials along these dimensions. To what extent *are* materials available, having the district's desired teaching approaches and type of curriculum content? If such materials as the school district desires are not available, what are the alternative approaches the district can consider? A school district looking for appropriate science materials at any level is faced with several options, among them the following.

Purchase materials from commercial publishers. The option most commonly selected by school districts is to purchase a textbook series from a commercial publisher. It is the most convenient option because of the easy availability of materials. The difficulty, however, is the lack of variety of teaching approaches and curriculum content. Similarities among competing books are much greater than their differences; market forces and school district preferences have led to remarkable similarity. Fortunately, some change is under way and appears to be continuing. The previously mentioned new family of NSF-funded elementary school programs is a clear example. More recently, a new family of middle school science programs was released after a similar development process. Both of these families of materials clearly are closer to the new National Science Education Standards than the materials in common usage. The greater degree of choice available in secondary science—particularly with respect to content concerning science/technology/society—is evident in *ChemCom* and the analogous programs under development in physics and biology. In addition to changes

Sidebar 4.4

NEW MIDDLE SCHOOL SCIENCE PROGRAMS

Recently a number of new middle school science programs have been developed, usually with the support of the National Science Foundation. Most of these programs are available from the publishers indicated.

Program and Developer	*Publisher*
Foundations and Challenges to Encourage Technology-based Science (FACETS) American Chemical Society 1155 16th St., N.W. Washington, DC 20036	Kendall/Hunt Publishing Co.
HumBio The Human Biology Middle Grades Curriculum Project Building 80 Stanford University Stanford, CA 94304-2160	Addison Wesley Publishing Co.
Insights: An Inquiry-Based Middle School Curriculum Educational Development Center 55 Chapel St. Newton, MA 02160	Developer
Introductory Physical Science (IPS) 6th Ed. Science Curriculum Inc. 24 Stone Rd. Belmont MA 02178-3521	Delta Education
Middle School Life Science (MSLS) Wheat Ridge Middle School 7101 W. 38th Wheat Ridge, CO 80033	Kendall/Hunt Publishing Co.
Middle School Science and Technology BSCS 5415 Mark Dabling Blvd. Colorado Springs, CO 80918-3842	Kendall/Hunt Publishing Co.
Science Education for Public Understanding Project (SEPUP) Lawrence Hall of Science University of California Berkeley, CA 94720	Addison Wesley Publishing

is course content, changes in the teaching approaches reflected in these materials also are visible.

Purchase materials developed by other districts. The last decade has seen an increase in the number of locally developed materials receiving widespread attention outside the confines of the school district where they were developed. A few such science programs are included in the materials available through the National Diffusion Network (NDN), a federally-sponsored endeavor for validating and making available programs developed by local school districts. The NDN processes give the purchaser assurance that the materials have been validated, and the description and claims attached to them are accurate.

Although they have not undergone the same extensive validation, there are many other locally developed science programs worthy of consideration. A good source for identifying them is the descriptions provided by the National Science Teachers Association (NSTA), as part of their *Focus on Excellence* series, initiated in the 1980s. Several monographs have been prepared which describe exemplary locally developed programs. The school district wishing to consider one of these programs is well advised to purchase these monographs from NSTA and examine closely the descriptions contained therein. Contact then can be made with the districts which prepared them to acquire more information and sample materials.

More recently, the NSTA initiated its Scope, Sequence and Coordination project,which stimulated numerous local curriculum projects. A number of these projects include the preparation of local instructional materials having a focus on integrating the science content of the various science disciplines. Some of these local projects are potential sources of integrated science teaching materials. Although in most cases the resulting materials are not developed sufficiently to be "exportable", they serve very well in the hands of people with a personal knowledge of their underlying rationale, idiosyncrasies, and role in the teaching practices of the teachers who developed them.

Although Project 2061 of the AAAS has yet to produce actual teaching materials, the design and development of curriculum materials is taking place at its test centers. In the future, some valuable teaching materials can be expected from these sites, in addition to the excellent theoretical publications this project has already developed.

Adapt existing materials. Another option for a school district is to select commercial materials—or those developed by other school systems—and then adapt them to fit the local situation. Such adaptation may require extensive curriculum development work, but there are concomitant advantages. One outcome of such an endeavor should be some degree of consensus among teachers as to how these materials should be used. In addition, such an endeavor should yield some type of guide teachers can use in adapting the purchased materials to the local situation.

Develop materials locally. Another option is for a school district to develop its own materials, possibly including a student textbook. Obviously, such a venture cannot be considered by most districts because of their size. In the case of a few large school districts, however, this option is viable, and it should not be overlooked when the materials available from other sources, such as commercial publishers, fail to meet school district needs. An extensive and careful study may be needed to determine whether such a venture is viable, but if the desired materials are not available, local development may be a valid option for large districts.

While such development endeavors usually are limited to quite large districts, occasionally a relatively small school district happens to have particularly talented teachers with the interest and energy to take this road. A review of the programs contained in the NSTA publications from their *Search for Excellence* illustrates this point. The district that has thoughts of moving in this direction may find such examples to be of help. In addition, conversations with experienced people may be particularly valuable in identifying the potential of this approach in a given context.

How to Develop Your Own Curriculum Materials

Even though most schools will find adapting materials from other schools to be more appropriate than developing their own, we will describe here a local school curriculum development process, both for its own sake and because the process of adapting materials may employ aspects of the development process described here. This development process contains six major phases consisting of creating, writing, pilot teaching, rewriting, field testing and production. The editing process is not considered a separate phase but is found within three phases—the writing, rewriting and production phases. It is not the only process that could be used, but it is presented here as one that has been found successful on several occasions. We recommend it to you.

Creating

The creating phase consists of developing a clear goal or purpose for the unit, and defining the audience; e.g., who are the students that will be using this material, what have they studied previously, are we preparing them for anything that follows this unit, what kind of students are they, and what are their interests? The second part of this phase consists of pulling together a group of people in a design conference. This group, which can vary in size from four to twenty four, should consist of (1) the small group of people who will actually do the writing after the design conference is over, (2) two or three experts on the topic of the unit, and (3) other generally creative people who know something about the topic and the audience.

This group should have a one and a half to three days of concentrated time together in which to listen to presentations on the unit content, participate in discussions on how students learn this content, and hold a series of brainstorming activities on the actual content activities and approach that will be created for student use. The salient feature of this particular conference is the mix of people who are brought together on a concentrated basis to create an outline for the unit. Any one category of people probably would be incapable of designing a creative unit, but together their experience, resources, and interactions typically produce a high quality, useful product.

One additional step is necessary before the writing team can begin the process of actually developing the unit. Assuming that the design team has produced a working outline for the unit, it is necessary for the program developers—together with members of the design team and writing team—to decide on the teaching/learning model that will be used in teaching the unit, and in the actual physical formatting of the materials.

By "teaching/learning model" we are referring to the beliefs or assumptions about the teaching and learning process that are held by the writers as the activities and other materials are written. Some models may be very simple. "Do the experiment before the lecture and explanation"; or the reverse. Deciding how and when to present vocabulary can be an important part of the teaching/learning model.

A model that the authors are very familiar with and one that has a fair amount of research behind it is the "learning cycle" introduced in the first edition of the Science Curriculum Improvement Study (SCIS), an elementary project at the University of California supported by the National Science Foundation in the mid 1960's.

The learning cycle consists of three phases: (1) exploration, (2) concept introduction and (3) concept formation. Recently, the BSCS Elementary Science program (1990) modified the learning cycle to include five phases: engagement, exploration, explanation, elaboration, and evaluation. Figure 4.1 provides a brief summary of these five phases.

Figure 4.1

The following is an overview of the teaching model used in *Science for Life and Living,* an elementary school science program developed by the Biological Sciences Curriculum Study.

Engagement
These activities mentally engage the student with an event or a question. Engagement activities help the students to make connections with what they already know and can do.

Exploration
The students work with each other, explore ideas together, and acquire a common base of experience, usually through hands-on activities. Under the guidance of the teacher, they clarify their understanding of major concepts and skills.

Explanation
The students explain their understanding of the concepts and processes they are learning. The teacher clarifies their understanding and introduces and defines new concepts and skills.

Elaboration
During these activities, the students apply what they have learned to new situations, and they build on their understanding of concepts. They use these new experiences to extend their knowledge and skills.

Evaluation
The students assess their own knowledge, skills, and abilities. These activities also focus on outcomes that a teacher can use to evaluate a student's progress.

A model, such as the above, is needed for organizing the activities to be presented to teachers and students. Once such a pattern has been established, a team of people can write the materials with a consistent pattern of work from the various people involved.

Do not overlook the importance of writing sample materials as an example of the format that will be used. If the format is well established before the total writing team begins their work, writers will be saved many hours of questioning and debate.

Writing

In the writing phase, a small group of people—usually two to four—spends a considerable amount of time together, time that should be concentrated in blocks several days in length. Unless the individuals in this group are experienced writers, the temptation to send people off to write on their own should be avoided. The interaction and guidance provided by a leader and the opportunity to interact far outweigh the advantage of people writing individually on their own schedules. If the writing team is relatively inexperienced, they should be monitored frequently during the early stages of writing to see that they are following the format and the teaching model, and that the objectives developed in the design conference are being addressed by the individual activities.

If both student materials and teacher guide materials are to be developed for the program, experience indicates it is more productive

---

From Bybee, R.W. and Landes, N.M.(1990). Science for life and living: An elementary school science program from Biological Sciences Curriculum Study. *The American Biology Teacher, 52* (2), 92-98.

to write student materials first and then later fill in the necessary instructions, hints, and supplementary material needed by the teacher.

One of the hardest adjustments for inexperienced curriculum writers to make is to accept and understand the fact that they are writing for other teachers, who will pick up the materials and use them without the advantage of extensive discussions with the author. Many teachers are accustomed to writing their own lesson plans, activity outlines, and worksheets, but have not had the responsibility of writing a stand-alone document that others must understand. The writers must learn to put themselves in the shoes of a teacher inexperienced with this particular unit or activity and tell that teacher—and sometimes the students—everything they need to know in order to successfully accomplish the activity or understand the concept being presented.

The writing team should consistently review each others' work, criticize it, and assist in any rewriting required to improve the material.

Trial Teaching

Once the material has been written and printed in a preliminary format, plans should be finalized for pilot testing the materials with a small group of experienced teachers. From our experience, the best group of teachers to pilot the materials are those who wrote the materials. A few additional highly competent teachers could be added to the original writing team if they have close contact with and support from the writing team. The purpose of this first round of trial teaching is to determine if the activities, ideas, and concepts work with the target audience of students. Therefore, it is important that the teachers involved know the new materials very well and be highly skilled in teaching them. If concepts are not understood and activities do not seem to work well, the problem can be attributed to the fact that they are inappropriate for the target audience. The developer should have enough confidence in the teachers that negative results in the pilot classrooms cannot be attributed to a lack of teacher understanding of the activity or a poor teacher presentation.

During the pilot process feedback can be collected in a variety of ways. The first method suggested is to develop a simple form for each activity for pilot teachers to complete at the end of that activity. Figure 4.2 is an example of such a form.

Figure 4.2. A Materials Testing Feedback Form

Activity Feedback

Name ____________________

Activity No. ________________

Time Required ______________

1. The percentage of students who met the activity objectives was

   ___90-100% ___80-89% ___70-79% ___50-69% ___less than 50%

2. Student interest in this activity was

Very High	5	4	3	2	1	Very Low

3. Students found this activity

Too Difficult	5	4	3	2	1	Too Easy

4. Were any student analysis questions unclear or unnecessary? If yes, which questions need to be deleted or modified?

5. Were the teacher pages clear and workable? If no, what changes need to be made?

6. If you were going to teach this activity again next week, what would you do differently?

7. What supplementary teacher information would be helpful when teaching this activity?

Comments:

Please return this form to the Science Department.

In some projects teachers have been asked to keep a daily log of what they do. This document shows their ideas and thoughts at the time they occur and also provides the developers with data on how long each activity lasts, as well as what other activities, procedures, films, etc. teachers use. Feedback forms such as Figure 4.2 will not pick up the supplementary activities that teachers often add as they go through a new unit.

It is wise to keep a record of the feedback as it comes in and promptly call upon teachers who are behind in submitting feedback. It takes discipline to complete the feedback form after every activity and send it in on schedule.

At regular intervals throughout the course, such as once a chapter, or unit, or month, many developers also ask teachers to fill out overall ratings of the past unit or month of study. The format is usually checklists or Likert scale items that are easily summarized. Such data are useful in compiling an overall evaluation of the project materials and are beneficial in making presentations to curriculum groups, district administrators, the public and others.

In many respects the most useful form of feedback can be obtained from a small group of teachers who have just finished a unit or several weeks of work and are gathered around a table to discuss with the developers their ideas on how to improve the materials. It is wise to keep this group small, pick them carefully, and listen to them closely. It is often useful to have pilot teacher records available at the meeting and refer to them as you go through the materials unit by unit. Keep a record of the decisions that are made about how to improve, rewrite, or modify the activities.

Do not forget to collect some student data in the form of questionnaires or interviews. Many evaluation or decision making groups will want to know what students think about the materials.

Rewrite

Some of the same people who sat around the table in the feedback session described above can be engaged to rewrite the materials based upon the feedback. The basic purpose of the rewrite is to refine, improve and probably reduce the amount of material available. New ideas can be considered at this point, but keep in mind that they will have less trial use and feedback than material that was in the original draft. It is also well to remember that you are writing for all teachers, not just the group of exceptionally competent and carefully selected teachers who were involved in the pilot test.

Field Test

The purpose of the field test is different than that of the pilot test. After materials have been refined, based upon the feedback on the pilot group, they should be ready for use by the typical teacher. To test this premise, randomly select a group of teachers who are representative of all teachers in the system. Make a conscious effort to find a diverse, typical collection of teachers, or select all teachers from some predesignated area of the district or collection of schools. The purpose of the field test is to determine if the materials can be used by all teachers and if the implementation support plan and in-service process are adequate to prepare teachers to use the new materials.

One way to see the difference between the field test and the pilot test is through the concept of controlled variables. In the pilot test the teacher variable was controlled by selecting a few highly competent teachers. The independent variable was the materials themselves. Would they work? In the case of the field test the materials are the controlled variable and the independent variable is the typical average teacher. What support will they need to make the materials work with their students?

Rewrite

This final rewrite provides the last chance to revise and improve the materials. Don't get creative at this point. Any rewritten or new materials cannot be tested beyond this point, so you are taking a risk by introducing new ideas and activities at this time. If all has gone well, this revision should be fairly limited and straightforward.

Materials Production

The final product should be attractive to teachers—and students, if student materials are required. A small extra effort and additional cost can result in very attractive looking materials. The availability of desktop publishing software makes the production of professional looking materials by the experienced user relatively easy. Purchase an attractive looking plastic three-ring binder or other cover and find an artist to illustrate the cover and portions of the guide if necessary. Do not overlook the possibility of having your materials bound by a commercial bindery; it may not be as expensive as you believe. You have put a lot of effort into developing the materials up to this point over the past two-to-three years. Go the one final step and produce an attractive looking product.

Adoption, Adaptation or Development Revisited

Having described a materials development process, consideration is given to the question of whether to adopt materials developed elsewhere, adapt such materials to fit your local situation; or develop your own. Although all three have their place, it may be that adaptation is most often the best approach to take because development is such a big endeavor and simple adoption too often does not include sufficient consideration of how the materials will be used to reach the new vision of science education you have in mind.

Adaptation, of course, covers a wide spectrum from the slightest modification of an adopted program to modifications so extensive that the beginning point is hard to recognize. The variations of such a process are so many it is hard to describe something typical, but it is probably safe to say that the more extensive the adaptation, the more it is likely to look like the development process just described.

For the sake of a tangible example, the work of a high school science department will be described here briefly. This example leans strongly to the development end of the adaptation spectrum just described, in that it pertains to a department that decided to initiate an integrated science program that extended over three years and replaced the biology, chemistry, physics, and earth science these students otherwise could take. The department did not get new textbooks in any of these subjects, nor did they have textbooks that integrated the science; they put together their own program using the textbooks already on hand. In addition, it was not just a new arrangement of the content; they were committed to a new vision of science education that included substantially more laboratory work, student involvement, student construction of understanding, and applications of science knowledge. Since the program had a totally new configuration, and three different teachers were involved in teaching this new science at each of the first two levels—and seven teachers over the three grade levels of the program—there was much that had to be done in common. As a result, much of what they did resembles the development process described above.

It was similar to the development model in that they selected learning activities, e.g., laboratory exercises and class activities, to make up the instructional program. They tested the activities in practice during the first year of use—a pilot test of sorts—prior to modifying them for use the second year. They used feedback from their first year as a basis of this revision. The process was long-term; they initiated the program one grade level per year. The "vision" for the program came from national recommendations and was translated into a program.

In other ways their approach differed from the above development model. They met weekly during the school year to plan for the next week and make group decisions about what should be done. As teachers they had no group "above them" doing any of the development work for them. Leadership for the program was of great importance, but it came from a teacher who also served as department chair, not someone outside the teaching ranks.

Such endeavors are not routine practice in the schools, but they may be more common than one would guess, either as entire departments or some subset of a department. Even so, the obvious question is how feasible such extraordinary efforts are, for they truly do involve a large amount of work. In our experience, the following ingredients are highly important:

1. A supportive climate in the school and a principal who truly values such endeavors,
2. Leadership for the project on the part of someone such as a department chair who is deeply involved, has the necessary leadership ability and commits the required effort to the task,

3. Collaboration on the part of the teachers that gets the work done, stimulates their intellectual efforts and provides the many needed forms of mutual support, and
4. The needed financial resources—often not large, but available when needed.

While all four of these ingredients may not be fully available, it is highly unlikely that such a venture can succeed in the absence of any one of the four. While all are essential, our experience shows that the power that really drives such an undertaking is the collaboration of the teachers. Collegial endeavors have great potential.

Next Steps

The acquisition of quality materials that reflect the ideal state of science education you are seeking is only one part of a long-term and involved process. Thus far, little has been said of how to use these materials in the process of helping teachers to utilize the best instructional strategies. It is hoped that the materials will foster the desired form of teaching and learning, or at least be compatible with them. In the next chapter, more attention will be given to processes of fostering this desired form of instruction.

References

American Association for the Advancement of Science. (1970) *Science a Process Approach.* Washington, DC: publisher. (AAAS)

American Chemical Society (1988). *ChemCom: Chemistry in the Community.* Dubuque, IA: Kendall Hunt.

Bredderman, T., (1984). Laboratory programs for elementary school science: A meta-analysis of effects on learning. *Science Education, 69*(4):577-91.

Education Development Center, (1970). *Elementary Science Study.* Manchester, MO: McGraw Hill.

Goodlad, John I. (1983). *A Place Called School: Prospects For the Future,* St. Louis: McGraw-Hill.

Goodlad, John I. (1983b). A study of schooling: Some findings and hypotheses. *Phi Delta Kappan, 64*(7), 465-470.

Goodlad, John I. (1983c). A study of schooling: Some implications for school improvement. *Phi Delta Kappan, 64*(8), 552-558.

Harms, Norris C., et al. (1981). *What Research Says to the Science Teacher,* Vol. 3, Washington, DC: National Science Teachers Association.

Johnson, D. & Johnson, R., (1987). *A Meta-analysis of Cooperative, Competitive and Individualistic Goal Structures,* Hillsdale, NJ: Lawrence Erlbaum.

Lawrence Hall of Science, (1978) *Science Curriculum Improvement Study II.* Boston: American Science and Engineering.

Penick, J.E., (1982). *Focus on Excellence: Science as Inquiry. 1*(1). Washington, DC: National Science Teachers Association.

Penick, J.E., (1983). *Focus on Excellence: Elementary Science. 1*(2). Washington, DC: National Science Teachers Association.

Penick, J.E., (1983). *Focus on Excellence: Biology. 1*(3). Washington, DC: National Science Teachers Association.

Penick, J.E., (1984). *Focus on Excellence: Physical Science. 1*(4). Washington, DC: National Science Teachers Association.

Penick, J.E., (1984). *Focus on Excellence: Science/Technology/Society. 1*(5). Washington, DC: National Science Teachers Association.

Shymansky, J.A., Kyle, W.C. & Alport, J.E., (1983). The effects of new science curricula on student performance. *Journal of Research in Science Teaching, 20(5),* 387-404.

Slavin, R., (1989). Cooperative learning and student achievement. In Slavin, R. (Ed.), *School and Classroom Organization.* pp. 129-156. Hillsdale, NJ: Lawrence Erlbaum.

Stake, R. and Easley, J. (1978). *Case Studies in Science Education,* Urbana, IL: The University of Illinois.

5

Fostering Improved Teaching

If new, top-quality curriculum materials are introduced *and* used in the intended manner, a major step has been taken in the process of educational improvement. This step, however, does not necessarily ensure first rate student learning. The quality of teaching is a central matter, and new materials do not always improve teaching.

An intensive study of a district-wide, quality, hands-on, elementary school science program focused in considerable detail on two schools in the district. In both schools hands-on materials were being used to do the activities described in the teacher's guide. Even though most indicators showed teachers in both schools were involving students in the activities as intended, the observers who spent many days in the two schools found some striking differences.

Most of the teachers in one school were teaching science very differently from the typical teaching approach found in the other school. In one school the students were allowed considerable freedom to explore with the materials, drew their own conclusions, were encouraged to view science both as a way of knowing and as a body of knowledge, and received grades based on the teachers' informal observations. In the other school, however, the students did the hands-on activities after receiving considerable background information and very detailed directions from the teacher. Vocabulary drills and considerable attention to remembering many facts were typical of the classes. The teachers developed their own paper-and-pencil tests even though their curriculum materials had none, and the district did not have a district science test. Students who followed directions and remembered many facts got the highest grades.

Both schools were doing hands-on science—a major departure from typical school practice—yet there was a great difference between the science teaching in the two schools. By many indicators, both schools had implemented the intended district science curriculum, yet student learning in the two schools was very different. Teachers in both schools saw hands-on science as the norm for their school and district, but did not realize how different their way of teaching science was from some other schools.

Some may argue that these differences are not really that important. After all, the students in both schools are doing hands-on science. But these differences can not be dismissed so lightly. In one instance specific facts are a central goal of instruction; in the other, they are learned—probably as thoroughly—as an ancillary outcome of instruction focused on the processes of investigation. Although students in both classes had the fun of working with the materials, the additional challenge in one instance was that of drill activities; in the other case it was the challenge of trying to answer science questions based on manipulation of and thinking about the materials in hand. Our "desired state" of science education described in Chapter 2 clearly is characterized by the latter of the two schools.

There also are potentially large differences in the character of the science knowledge *per se* students learn in each instance. Learning more facts is not the same as acquiring an understanding of a fundamental science concept. Engaging in activities involving data acquired from thermometers, for example, may provide a knowledge of temperature scales, or it may lead to an understanding of the difference between temperature and quantity of heat. The latter conception, of course, is one that generally takes considerable class time and experience to acquire.

It probably is apparent that the approach favored here is the one presented in Chapter 2 under the label of "constructivist" learning. As described there, it is grounded in four principles.

First, learning is a process of students constructing their own meaning. Second, learning depends upon the preconceptions students bring to a subject, i.e., meanings they have already constructed at a prior time. Third, learning is dependent upon the context in which the concepts are encountered. Fourth, meaning is socially constructed; understanding develops through interaction between student and teacher and between the student and other students.

Sidebar 5.1

Fourth grade children in an elementary school class considered what would happen if they placed a thermometer inside a sweater. Everyone in the class said that it would get hot; the thermometer would register a higher temperature. They were baffled when they actually tried it and found that the thermometer indicated the same temperature whether it was inside the sweater or not. Their convictions were strong; the evidence was not powerful enough to change their minds. They were convinced they needed to leave the thermometer in the sweater longer, or do something to be sure cold air somehow was not getting in there.

While the teacher had thought the one experience with the thermometer in the sweater would settle the matter and the class would be able to move on, it did not turn out that way. The children had run up against a situation in which *their concepts* of heat and temperature did not correspond to *the natural world.* Their concepts were deeply entrenched. Simply telling them the "correct answer" had essentially no chance of changing these basic conceptions. After three days of conducting various tests in their classroom, some of the students still had not abandoned their personal conceptions in favor of "correct" concepts of heat and temperature. Their personal experience with sweaters "making them warm" stood in the way of acquiring new concepts. Three days of additional experiences in the classroom led to some new conceptions for many of the children. Some progress had been made, but even the best of "hands-on, minds-on" education would not bring all of the children to the point the teacher desired. Basic science conceptions are difficult to change. Just telling them, in a case such as this one, has little hope. With well developed instruction, much more progress can be made, but it still is not easy.

These experiences in an actual classroom are the basis for an excellent article about children's learning in science by Bruce Watson and Richard Konicek in *Phi Delta Kappan*. The reader wishing further insight to the process of teaching fundamental concepts in science will find the article worth reading: Watson, B. and R. Konicek. (1990). Teaching for conceptual change: Confronting children's experience. *Phi Delta Kappan, 71*(9): 680-685.

Instruction that fosters authentic student learning can take many forms, but whatever its form it helps students (1) connect the new understanding with prior knowledge, (2) check for inconsistencies with this prior knowledge, (3) alter understandings as needed, and (4) test new understandings in yet additional contexts. It is an activity for which students must take responsibility; the teacher is there in the role of coach to help the students in this endeavor.

Although it may not be given the "constructivist" label we have used here, the fundamental principles of learning and teaching under discussion here are increasingly seen in the work of science education reform groups, be they curriculum development organizations, professional organizations, or standards setting groups. An example is the set of principles of learning adopted by Project 2061 in *Science for All Americans* (see sidebar 5.2).

New Instructional Approaches

Translating principles of learning such as we have been discussing into specific teaching approaches is no small task. Increasingly, however, it is the subject of research and practical materials prepared for teachers. James Minstrell, for example, an accomplished high school physics teacher and educational researcher, has published the results of several investigations of teaching approaches having these characteristics. An increasing number of articles written for teachers are appearing in periodicals.

To a large extent it is not the specific teaching strategy so much as it is the manner in which it is used, but certain strategies clearly have more potential than others. Some of the strategies often advocated today have especially high potential in this regard. Examples are offered here of approaches with the potential of being used in a constructivist manner.

Cooperative learning

With its dependence upon group activities and cooperation among students to accomplish learning goals, cooperative learning generally takes on quite a different form than conventional instruction.

Sidebar 5.2

The report of the American Association for the Advancement of Science's Project 2061, *Science for All Americans* (1989), states their principles of learning as follows.

Learning Is Not Necessarily an Outcome of Teaching

Cognitive research is revealing that even with what is taken to be good instruction, many students, including academically talented ones, understand less than we think they do....

What Students Learn Is Influenced by Their Existing Ideas

People have to construct their own meaning, regardless of how clearly teachers or books tell them things. Mostly, a person does this by connecting new information and concepts to what he or she already believes....

Progress in Learning Is Usually From the Concrete to the Abstract

Young people can learn most readily about things that are tangible and directly accessible to their senses—visual, auditory, tactile, and kinesthetic....

People Learn to Do Well Only What They Practice Doing

If students are expected to apply ideas in novel situations, then they must practice applying them in novel situations....

Effective Learning by Students Requires Feedback

The mere repetition of tasks by students—whether manual or intellectual—is unlikely to lead to improved skills or keener insights. Learning often takes place best when students have opportunities to express ideas and get feedback from their peers...

Expectations Affect Performance

Students respond to their own expectations of what they can and cannot learn. If they believe they are able to learn something, whether solving equations or riding a bicycle, they usually make headway. But when they lack confidence, learning eludes them....

From *Science for All Americans* (pp 145-147) by American Association for the Advancement of Science 1989, Washington, DC: American Association for the Advancement of Science.

Research (Johnson & Johnson, 1987, Slavin, 1989) clearly documents its potential for increased learning, particularly with respect to some of the important goals of science instruction. Properly done, cooperative learning can be used to foster student investigations and interactions about ideas in a manner that will foster learning in the constructivist sense.

Simulations and role-playing

A simulated city council meeting—and the preparations leading up to it—contained in the first module of the *ChemCom* high school chemistry course serves as a good example of the use of a simulation activity in science instruction. It involves the students in developing alternative positions with respect to the competition between environmental and economic interests when a chemical plant creates a pollution problem in the community where it is the largest employer. As in the case of cooperative learning, such a simulation activity may provide a context in which student construction of important understandings can occur. If the goals of science instruction are to include applying science knowledge to science-related societal issues, students must have the opportunity to address these matters directly. Simulations, role-playing and related approaches have the potential for providing the needed opportunities for students to construct meaning and new understandings with respect to these issues.

Student projects

Student projects of many kinds—not just the science fair variety—can be used to foster student investigations and help students gain an understanding of how science investigations proceed and stimulate learning of important science concepts. But how they are organized and managed by the teacher is crucial to what outcomes actually result. They can be routine activities in which students are adhering to a lock-step formula, or they can be stimulating activities with a high level of engagement in learning. If students are given the opportunity to formulate their own questions to be addressed, design their own investigations, and work independently to find answers to the questions, there is real hope that they will come to understand the nature of scientific investigations and their related outcomes.

The key to successful use of these approaches is the competence and initiative of the teacher. To a large extent the performance of the teacher will depend on what he or she brings to the situation when initially employed in the schools. But it does not end there. Good teachers continue to grow throughout their careers. Much of this professional growth depends upon the drive, interest, and skill of the individual teacher. There is much that leaders in schools and school districts can do to facilitate this growth.

Fostering Quality Teaching

Helping teachers acquire competency in using a new instructional approach is a major undertaking. Research shows that it takes 45 to 65 hours of instruction, practice and coaching for teachers to acquire a high level of proficiency with only a moderately difficult teaching strategy (Shalaway, cited in Garmston, 1987). It may require that teachers reexamine some of their values about students and learning. It may require that they examine some of their beliefs to see if they are consistent with the results of research on teaching and learning. It certainly will require an opportunity to learn and practice new skills needed in the classroom setting. Such changes in behavior do not occur easily; they are a result of helping teachers reconstruct their understanding and methods of teaching.

Although numerous types of activities are used to foster the improvement of instruction, the discussion here will be limited to four broad categories of activities: in-service education classes, clinical supervision, peer coaching, and intensive programs of instructional improvement. Given the breadth of these four groups, they include a substantial proportion of the teaching improvement endeavors currently in use.

In-service education classes

In-service education classes can take many forms. In addition to those focused on the appropriate use of new curricula, such classes can address specific teaching approaches such as cooperative learning, inquiry laboratory instruction, or general teaching topics such as classroom management. They can be individualized programs, designed to address whatever matters an individual wants to pursue. They can range from courses which are very instructor-directed to those in which the members of the class largely take responsibility for their own learning.

In their discourse on in-service education, Sparks and Loucks-Horsley (1990) address five types of staff development programs:

Individually guided staff development — a form of staff development in which teachers individually identify what they think will improve their teaching and then plan for and pursue their own plan for growth and improvement.

Observation/assessment — an approach based on someone observing a person's teaching and then providing feedback to the teacher concerning the classroom performance.

Development/improvement process — professional growth results from working in a program to develop new curricula, design a new program, or engaging in a school improvement program addressing instructional problems.

Training — individual or group instruction designed to help the participants acquire new knowledge or skills. (What many people think of when they encounter the label, "staff development.")

Inquiry — an approach in which teachers identify an instructional area of interest and then engage in a process of collecting data and making desired changes in this instruction based on these data. (pages 234–250).

All of these approaches can be used for the full range of purposes—including both general and specific science education topics as described above.

Sidebar 5.3

In a meta-analysis of research on science teacher education, Yeany & Padilla (1986), synthesized data from a substantial number of research studies that demonstrated the increased effectiveness of a combination of techniques; it is presented in a graph that is reproduced here.

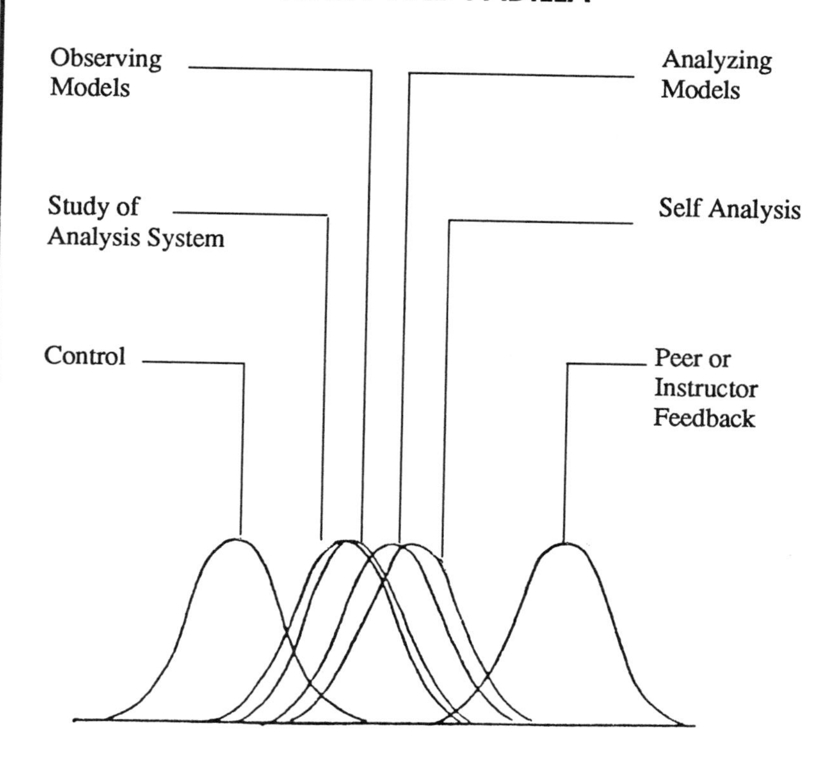

Theoretical distribution of effect sizes for five types of strategy analysis training.

From Yeany, R.H. & Padilla, M.J. (1986). Teaching science teachers to utilize better teaching strategies: A research synthesis. *Journal of Research in Science Teaching*, 23(2): 85–96.

Two types of in-service education in particular are of interest here because of their wide applicability to science education improvement: in-service classes for the implementation of new curricula and classes designed for helping teachers initiate a new instructional approach. The first of these two, in-service classes for curriculum implementation, will be addressed in the next chapter. The second, learning a new instructional approach, can stand on its own as the focus on an in-service education endeavor, but our experience indicates it will be more effective if it is part of a broader curriculum implementation or improvement endeavor. In either case, this type of in-service education program is a *major* undertaking, requiring an extensive amount of time and opportunities to practice the new instructional approach.

Research also indicates that proficiency in using a new strategy increases as various modes of instruction are added to the in-service education endeavor. The teacher who receives only ordinary in-service classroom instruction on the use of the new teaching approach will have a level of proficiency quite unlike the teacher who, in addition, has an opportunity to see the approach modeled, to practice it, get feedback on this practice activity, and finally try the approach again with the benefit of insights gained from this feedback.

Teachers need opportunities of several types to improve their teaching practices—in-service education and a variety of others. Here are some guidelines for the four categories identified earlier: in-service education classes, clinical supervision, peer coaching, and instructional improvement programs.

In-service Education. Some general guidelines for effective in-service instruction are offered here.

1. Fully define, in operational terms, the instructional approach that is to be taught. Models of it should be available in both videotape and written form, and persons should be available who can demonstrate it in classroom settings.
2. Develop a class of sufficient length and intensity that the participants have a realistic opportunity to learn the new approach.

3. Provide opportunities for the participants to practice the new approach under circumstances where they can receive feedback on what they do, e.g., through viewing videotapes of what they have done and interaction with an instructor or a peer who can assist in the personal assessment process.

Well developed in-service education that will make a major impact on *how* teachers teach clearly is not a minor undertaking. It is a topic, however, that extends beyond the scope of this book. For more in-depth information, the reader is referred to Joyce & Weil, 1986; Sparks & Loucks-Horsley, 1990; and Joyce & Showers, 1983.

Some forms of assistance described above as a part of effective in-service education classes cannot be provided for teachers easily in the context of classes alone. Related activities such as the following have the potential of substantially enhancing the process of helping teachers acquire new teaching competencies.

Clinical Supervision. The traditional role of a supervisor has included providing feedback and helping the persons under his or her supervision to improve their work. In reality, of course, the evaluative portion of the role has overshadowed this improvement process, and its value is often questioned by teachers. Under the right circumstances, however, a skilled mentor should be in a position to provide more intensive and effective assistance than would be possible through an in-service education class.

Clinical supervision, a label attached to a variation of the supervisory process, is founded on two basic premises. First, the supervisor will record only objective information during an observation, not information of an evaluative nature. Second, during the subsequent feedback session, the supervisor will report only this objective information and offer no critique of the teaching. Only open-ended questions will be raised as a point of discussion. The teacher who is being observed is the only person who can direct the conversation into an evaluative mode.

One of the sources of the power of the process comes from the documented reality base that the supervisor brings to the process through records kept of the classroom activities. This person is expected to take notes—verbatim to the largest extent possible—and provide this information to the person observed. It is often termed scripting, in that the intent is to provide a script of the class as it actually occurred. Although it is a skill that needs to be developed, many supervisors have found they have been able to develop the ability to take detailed notes of most of what is occurring in the room, including actual statements of teacher and students.

Another source of power in the process is the analysis and reflection that the teacher under supervision brings to it. With the script of the class at hand, the teacher is in a position to address reality—not what his or her memory says happened—and reflect on what occurred, consider alternative actions that may have had a different result, and explore questions both philosophical and pragmatic. The process is very consistent with a constructivist view of learning, in that the teacher must construct his or her own meaning and it is done in the context of actual teaching.

In its purest form, this type of supervision would focus on whatever agenda the teacher wished to pursue. It can, of course, be used in a variety of ways, including in the context of implementing a change in curriculum.

Peer Coaching. The benefits of a mentor, confidant, or sounding board are not just available from someone higher in the school hierarchy. Under the right circumstances they may be available from a peer as well; peer coaching is becoming more common. As in the case of assistance from supervisors, it has varying philosophical orientations with varying degrees of directiveness from one's peers.

Garmston (1987) describes three types of peer coaching: technical, collegial, and challenge coaching. He uses the term technical coaching with reference to coaching done as part of a well-developed program of transferring training into classroom practice. In other words, it is an adjunct to an in-service program of training in the use of a new teaching approach. Although done by peers, i.e., fellow teachers, it is an integral part of a lengthy, 2-3 week, full-time equivalent, in-service program in which the participants are acquiring a new instructional approach. It obviously is costly and may have some tendency to inhibit collegiality, a core goal of the next type of peer coaching.

The second type, collegial coaching, is intended to refine and polish teaching practices without being limited to a particular teaching approach in an in-service program. It is intended to foster collegiality, promote professional dialogue, and help teachers reflect on their professional work. The observed teacher's personal priorities set the agenda, not an in-service workshop. As in the authentic clinical supervision described above, the role of the observing teacher is not to pass judgment, but to assist the observed teacher by collecting information, helping to analyze and interpret it, and serve as a sounding board in the reflection process. As in clinical supervision, a constructivist approach to learning prevails; teachers construct their own understanding in the context of their teaching environment.

A third version of peer coaching, challenge coaching, is focused on resolving a specific problem common to teaching in a particular setting. In contrast to the other two versions which most often are done in pairs, this version is more likely to be done in a group setting. The problem on which the endeavor is focused may be, for example, the difficulties of using cooperative learning in a new junior high science curriculum, using hands-on science materials in an elementary school, or doing inquiry-oriented laboratory work in a senior high school; or it may be limited to teaching some particularly difficult concept or concepts.

All three versions have the potential of breaking down the common debilitating isolation in which teachers work, improving teaching practice, and increasing teachers' sense of efficacy. To work effectively they demand knowledgeable leadership, resources, and a sensitivity to the school culture. In particular, attention must be given to defining the form and purpose and eliminating any hidden agenda on the part of administrators in what is billed as *peer coaching*.

Instructional Improvement Process. Another approach to the improvement of teaching may involve an entire school. Under such an instructional improvement process, a team of teachers and supervisors visits a school—or in some cases a department—as a team for several days—usually spread over a school year—to assist the teachers and administrator in reviewing their instruction and developing plans for its improvement. The curriculum can be a focus of attention as well as the instruction. For example, if the "key features" of the science curriculum have been clearly described, they can provide the basis for a review of current classroom practice in which actual practice and the ideal are compared. Based on an analysis of the curriculum and instruction, the team can go on to provide assistance in a manner that is mutually agreed upon by the team and the school staff. What results from the process can take many forms, including in-service education and one or more forms of supervision or coaching as described above.

The improvement of instruction is a large and important job. The allocation of substantial leadership and resources can pay big dividends in terms of student learning. It is an important component of a total systemic science education improvement endeavor.

References

American Association for the Advancement of Science. (1989). *Science for All Americans.* Washington, DC: American Association for the Advancement of Science.

Garmston, Robert J. (1987). How administrators support peer coaching. *Educational Leadership,*, 18-26.

Johnson, D. & Johnson, R. (1987). *A Meta-analysis of Cooperative, Competitive and Individualistic Goal Structures.*, Hillsdale, NJ: Lawrence Erlbaum.

Joyce, Bruce & Weil, Marsha. (1986). *Models of Teaching* (3rd ed.). Englewood Cliffs, NJ: Prentice-Hall.

Joyce, B.R., & Showers, B., (1983). *Power in Staff Development Through Research on Training.* Washington, DC: Association for Supervision and Curriculum Development.

Linn, Marcia C.. (1986). *Establishing a Research Base for Science Education: Challenges, Trends, and Recommendations,* Berkeley, CA: Lawrence Hall of Science.

Pines, A. Leon. (1985). Toward a taxonomy of conceptual relations and the implications for the evaluation of cognitive structures. In Leo H. T. West & A. Leon Pines (Eds.). *Cognitive Structure and Conceptual Change.* Orlando: Academic Press.

Shalaway, Linda. (1985) Peer Coaching ...does it work? *R & D Notes,* National Institute of Education, (September 1985): 6-7. In Robert J. Garmston, How administrators support peer coaching, *Educational Leadership,* February 1987.

Slavin, R., (1989). Cooperative learning and student achievement. In Slavin, R. (Ed.). *School and Classroom Organization* (pp 129-156). Hillsdale, NJ: Lawrence Erlbaum.

Sparks, D., & Loucks-Horsley, S., (1990). Models of staff development. In W.R. Houston, (Ed.). *Handbook of Research on Teacher Education* (pp 234-250). New York: Macmillan.

Watson, B. & Konicek, R. (1990). Teaching for conceptual change: Confronting children's experience. *Phi Delta Kappan, 71*(9), 680-685.

West, L.H.T., & Pines, A. L., (Eds.), (1985). *Cognitive Structure and Conceptual Change,* Orlando: Academic Press.

Wittrock, Merlin C. (1986) Student thought processes. In Merlin C. Wittrock (Ed.). *Handbook of Research on Teaching* (p 260) (3rd Ed.) New York: Macmillan Co.

Yeany, R.H. & Padilla, M.J. (1986). Teaching science teachers to utilize better teaching strategies: A research synthesis. *Journal of Research in Science Teaching,* 23(2): 85-96.

6

The Change Process: How to Implement and Maintain an Innovative Program

There are two major threads that can be followed in this book; the first is the issue of what to change and the second is how to go about changing it. The first thread, what to change has been highlighted in the chapters on goals, materials and teaching. The second thread, *how* to go about changing it, started in Chapter 1 with the overview of the phases of an improvement plan. In Chapter 3 the process of establishing a need became the first step in the process of change. Chapter 5 dealt with both the ideal instructional methods and strategies and the process of helping teachers make the necessary changes. This chapter will take a serious look at the process of change and provide specific suggestions on how to make and maintain improvements through change.

Michael Fullan (1982 and with Steigelbauer, 1991) points out that the change process involves changing materials and changing peoples' beliefs and behaviors in the use of those materials. All three aspects of change are necessary to achieve a particular educational goal, but often individuals change one or two of these dimensions without affecting the others. A teacher could use a new curriculum material or technology without altering his/her teaching approach, or another teacher could use materials and alter some teaching behaviors without coming to grips with the concepts or beliefs underlying the change.

Just as constructivism is beginning to have an impact on the way we view the learning of students, it is becoming a strong influence in understanding how adults learn and how they change their beliefs and behaviors (Hutchinson & Huberman, 1993). From this perspective, knowledge and beliefs are largely self-constructed through social interaction. In classroom practice this results in what has been called "mutual adaptation," the process whereby innovations which were intended for use in a given manner are modified by the users to fit their self-constructed view and understanding of how to get the most success for their students (Fullan & Pomfret, 1977). As an example, the difference of just using an innovation in a mechanical way and using it with deep conviction and understanding is illustrated in the discussion in Chapter 5 of the two schools using the same elementary science program.

Much of the literature in science education over the past thirty years has dealt with change of materials, i.e., the development of PSSC Physics, ESS Elementary Science, ChemCom Chemistry etc., without the necessary and appropriate attention to helping teachers construct new approaches and beliefs required to get the desired effect from the new materials. Recently, attention has turned to viewing change beyond the individual and to thinking in terms of long term impact on the total system (Darling-Hammond, 1990). Furthermore, the systems are schools or school districts, and the evidence is clear that schools are cultures with their own rules, mores, and feedback mechanisms designed to resist change (Sarason, 1971). Although the systemic perspective is taken by this book (See Chapter 1) it is, nevertheless, useful to concentrate on individual change as an important and critical place to focus and start our discussion.

It may be helpful to use the metaphors of producer-consumer and collegiality when discussing the change process. The producer-consumer is a top-down process; someone outside the classroom makes a decision concerning changes that will occur in curriculum or teaching practices within the classroom. In other words, there are those (the producers) who promote improvement for the classroom and define what school personnel (the consumers) ought to be doing. There's a fairly strong dichotomy between the change agents and implementers who have the power and authority, as compared to the users or classroom teachers, who are perceived to be much less powerful. However, in reality the final power or ultimate gatekeeper in the process is the classroom teacher, who behind his or her closed door, makes a very personal decision about whether or how to use a particular innovation.

In the collegial approach, different groups of people are viewed as having different kinds of expertise and being involved at different times in the development of an innovation. From this perspective innovations are often interpreted and modified by the various players at all stages—from development to final classroom use. The producer/consumer distinction becomes blurred as various individuals influence others throughout the entire process, each contributing to the outcome without concern about their official role or position in the organization. In such a process products become very dynamic, being modified or self-constructed as they are developed and implemented. When the user modifies or adapts the innovation in the process of implementing it and the developer capitalizes on this change and builds it into the innovation; "mutual adaptation" has occurred. In this case it's hard to tell who the actual developer is. This is in contrast with the perspective that a "product" is introduced and handed down from the innovator to the consumer in the producer/consumer metaphor.

The research literature sheds some light on the efficacy in producing change of these two orientations. Researchers such as Olson (1980) and Aikenhead (1985) have produced a number of reports indicating that mandating new, innovative curriculum materials has little effect on teachers' classroom practices.

The collegial perspective reflects the assumption and evidence that people are willing and able to change what and how they teach, and are willing to be a part of an organized change process in which they can be assisted in making the change through in-service, administrative support and pressure, and other interactions that occur between the change facilitator and the individual classroom teacher. Some researchers have found that a behavior change often must occur first and that the belief follows the change in behavior. In other words, individuals act their way into a new way of thinking; not think (or believe) their way into a new way of acting.

Three research projects will shed light on the question of whether or not major implementation efforts can be successfully conducted with a large number of teachers and schools. One of the first major studies of implementation was the so-called Rand Change Study reported by Berman and McLaughlin (1977). Their research focused on the hundreds of local projects undertaken during the late 1960's and early 70's funded by the federal government (ESEA Title V), which provided millions of dollars to local school districts to develop and implement local innovations. The single most significant result of the study was that this national effort to sponsor local educational projects was a failure. In other words, after federal funding was withdrawn, usually three years after the initiation, most of the projects were non-existent, having lost local support when the federal resources were withdrawn. What Berman and McLaughlin did report was that in those rare instances where schools did succeed in maintaining the new programs, they were often modified by users in the process of implementing the innovation. This study tends to support the notion that the developer cannot expect such innovations to be implemented with a high level of fidelity, i.e., with few if any changes or modifications. But if success is defined as use with mutual adaptation, there is much greater chance that it will occur.

A few years later Crandall (1982) reported the results of the Study of Dissemination Efforts Supporting School Improvement (DESSI) with a much more positive finding. In their study of 144 selected schools, they found that change had taken place and improvement was occurring.

What could account for the significant difference between these two major studies only a few years apart? Unlike the Rand Studies, the DESSI group looked at schools supported by federal and state programs that had been "validated" by one of several federal agencies including, ESEA, National Diffusion Network, or the Office of Special Education. Such programs had been carefully developed, well defined, and determined to be effective by a set of evaluation criteria.

Although both of these studies were long, rich and complex, one outcome of the DESSI Study was the development of an "ideal scenario" that described the factors were most likely to lead to successful implementation. Such a scenario contains the following major ingredients:

The innovation fits the need.

The innovation is well defined.

There is strong administrative initiative and pressure on teachers to use the innovation, at least during the initial year.

A high level of technical and personal support is provided to all teachers.

The innovation is implemented, producing the desired classroom results.

Ongoing pressure and support are provided, resulting in a stabilized use or implementation of the innovation.

Although a number of other scenarios were also described in the research, this one produced by far the most positive outcomes. The general tenor of this book is that change can occur and needs to occur. Much of the discussion in this chapter and the balance of the book will draw from both the producer/consumer and the collegial perspectives in delineating a proactive approach to change. In this regard, the ideal scenario described in the DESSI study provides a very useful model to guide supervisors, department chairs, principals, and teachers as they plan and carry out innovations in their local schools. In the rest of our discussion we will use the term change facilitator to cover all the possible people who could be responsible for introducing an innovation and supporting its implementation. Nevertheless, the local context including the experience, attitudes, and responses of the individual teachers will be seriously considered throughout the process.

Does the Innovation Fit the Need?

A well documented and understood need for the innovation is a must! (An innovation can be anything new to a group of people. It can be a new organizational structure or decision making process in a school. A new curriculum is a frequent innovation in schools or districts. It can also be a new approach to instruction such as cooperative learning, teaching "thinking skills," wait time and other discussion techniques, etc.) If the need is understood, does the innovation fit the need? Is there evidence that it works or produces the needed results?

These are tough questions that must be answered by the change facilitator and the staff involved. The needs assessment process outlined in Chapter 3 will go a long way toward answering the first question. Developing and selecting an effective innovation that meets the desired state criteria described in Chapter 2 will do much to answer the latter questions. If the innovation was developed outside the local school district, check carefully with the developer to see what evaluation data are available. These data probably won't simply be the results of standardized tests demonstrating that the experimental approach is superior to the traditional or control group. Rarely are data of this type appropriate to answer the questions about the effectiveness of the innovation. Look instead for a wide variety of formative or summative data from the developer, describing how the innovation has met the objectives it set out to meet. This may be in terms of objective or criterion referenced test data, evaluation comments by professional evaluators, or documented observations by the pilot and field test teachers. In any event, check the data carefully and share them with a wide variety of teachers and administrators in your district. They probably represent only the first step in the process of determining whether or not the innovation meets the locally determined needs.

The next step will be the process of adaptation and construction of understanding by the local staff through your own pilot and field test. This process is very similar to the one described in Chapter 4. At the end of the field test year, data similar to those available from the developer, along with descriptions of the local adaptations, can be presented to the district curriculum committee, school board, or whoever must make the decision concerning final implementation of the innovation. At this point, whether the innovation fits the originally described need should be reasonable clear.

The Innovation is Well Defined

Before full scale implementation of the innovation, consider another important issue: Has the innovation or the change been well defined for everyone involved, particularly teachers and principals who must implement it? One of the fundamental reasons that many changes do not occur is that the participants involved do not understand the changes needed to produce the desired results, the reasons for the changes, and the latitude allowed in making the changes before they lose the desired effect.

Curriculum developers, and those who create innovative instructional strategies, are usually very careful to spell out the objectives for students so that certain behavioral outcomes can be measured. Why not do something similar for the teachers who must use the new teaching techniques or materials? The most obvious way to do this is to put explicit directions in the teacher's guide that describe behaviors and procedures teachers should use in conducting a lesson. Such a "road map" may look very prescriptive, but need not be if several alternatives are outlined so that teachers have choices at many points.

A major issue that the local improvement committee must resolve is how much the innovation can be modified and still be acceptable. If an innovation has been validated by the developer and verified through the pilot and field test process, you may be convinced that too much adaptation will destroy its effectiveness. Each committee or local school district will have its own position on this matter, but it is important to have a means of communicating the degree of adaptation that is acceptable. If a series of classroom observations were made across the district or school, one could expect to find a variety of patterns of how teachers have adapted or configured the innovation in their classrooms. To document ways in which teachers implement or adapt the innovation in their classroom, the Innovation Configuration was developed by Hall and Loucks (1978) as a part of the Concerns Based Adaptation Model.

The Innovation Configuration for an innovation can be thought of as a matrix with the program components such as use of materials, teacher role, and student activities, on one axis and two to five variations describing the different ways each component is demonstrated in the classroom on the other axis.

A simplified example of the matrix format for the Innovation Configuration is provided in Figure 6.1. The heavy dark line separates the "ideal" variations from the "non-ideal" variations. This enables the observer to compare classroom use of the innovation to "ideal" use. The reader will note that more than one variation may be "acceptable" though not "ideal." "Acceptable" variations are listed between the solid and broken lines. All practices to the right of the broken line are "not acceptable."

Figure 6.1. Science Program Innovation Configuration

IDEAL	ACCEPTABLE	NOT ACCEPTABLE		
Component 1: Units Taught (1) All units and most activities are taught	(2) Most units and activities are taught	(3) Some units are taught	(4) A few selected activities are taught	(5) No units or activities are taught
Component 2: Use of Materials (1) Students are constantly manipulating science materials	(2) Selected students only and the teacher handle the materials most of the time	(3) Typically, the teacher does demonstrations and the students watch	(4) No materials are manipulated by students or teachers	
Component 3: student Groupings (1) Students work individually and in small groups	(2) Students are kept in 3–5 permanent groups	(3) The whole class is taught as a group		
Component 4: Process/Content Emphasis (1) Science content and science processes are emphasized equally	(2) Science content is given major emphasis	(3) The processes of science are given major emphasis	(4) Memorization of facts and reading about science are emphasized	
Component 5: Assessment (1) All assessment activities are used	(2) Some assessment activities are used	(3) Teacher-made tests are used all of the time	(4) Learning outcomes are not assessed	

The scheme for developing an Innovation Configuration has been reported by Heck, Stiegelbauer, Hall and Loucks (1981) and is illustrated in Figure 6.2. This generalized process is designed to start with the developer's viewpoint on how each component might be used and expand or adapt it based on the agreed upon local decisions. The developer's viewpoint is usually a variation labeled "ideal." Other variations are designated as "acceptable" or "not acceptable." An Innovation Configuration should include all variations observers might actually see used in classrooms. In practice, both the components and variations for each must be held to a workable number, usually 8-10 components and 3-5 variations.

Figure 6.2
Procedure for Developing an Innovation Configuration

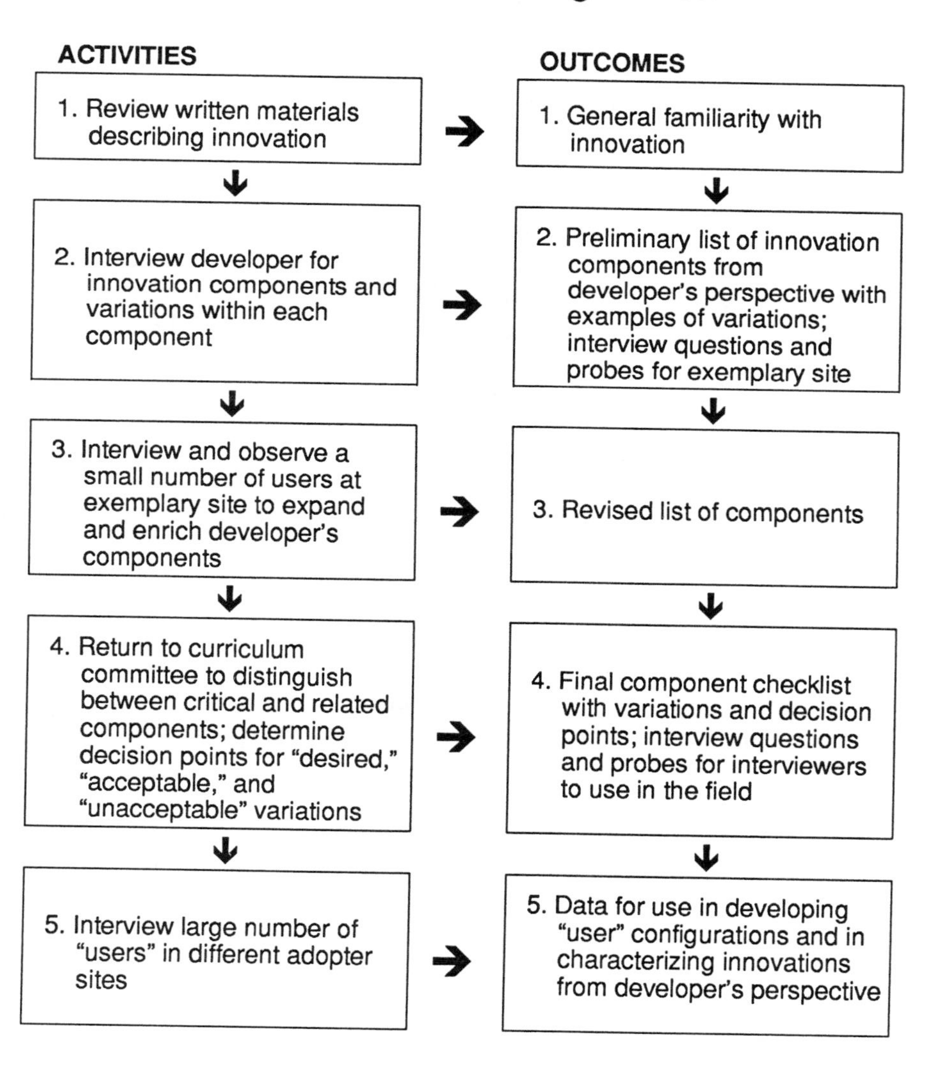

Adopted from S. Stiegelbauer, G.E. Hall and S.S. Loucks, (1981). *Measuring innovation configurations: Procedures and applications.* Austin, TX: Research and Development Center for Teacher Education, The University of Texas.

Uses of the Innovation Configuration

One way of using the Innovation Configuration is in one-to-one interviews of teachers, with the interviewer simply marking the variations the teacher reports. As an alternative, the matrix can be converted into a questionnaire, with the variations listed for each component and teachers asked to check the variation which most closely approximates their use.

1. *Choosing and Shaping Interventions.* Comparing responses to the relative values of the three kinds of variations ("ideal, "acceptable," and "not acceptable") in use will suggest the kinds of strategies required to support the implementation. For example, if teachers were demonstrating learning activities instead of allowing students to participate, and this was judged "not acceptable," then the change facilitator will want to investigate why and encourage the teacher to allow students to manipulate materials. That might be as simple as unlocking the equipment closet or as complex as helping the teacher learn to manage an activity-based program.
2. *Monitoring and Supporting Implementation.* Innovation Configuration data can be used to examine implementation on a classroom-by-classroom basis to discover predominant patterns (configurations) of use. The data can be sorted according to subgroups of individuals with similar patterns of use. For example, these results might show that all inexperienced teachers were having problems with the management of material, or that particular teachers were not involving students in the laboratory. A support system could be designed to meet the needs of each subgroup. The collection of data will make it obvious that not all teachers are using the innovation in the same way, nor do they all need the same type of assistance. The support system is designed to encourage mutual adaptation, with limits worked out by the local curriculum committee.
3. *Communicating to Teachers on How the Innovation is to be Implemented.* The Innovation Configuration can be used to encourage teachers' understanding and use of this innovation, while simultaneously communicating the school's or district's expectations about the implementation. Experience with numerous

implementation efforts shows that many teachers do not know how they are expected to implement a program.

4. *Setting Implementation Goals.* The Innovation Configuration can be used for classroom, school, or district goal setting. For example, it could be agreed that some components would be implemented first, and others later. Alternately, a district could stage the implementation across variations, with certain variations being labeled "acceptable" during the first year, but "not acceptable" during the second or third year in order to encourage the staff to implement in a more "ideal" manner. Thus, the Innovation Configuration can help in planning implementation and serve as a monitoring device for the process. Change facilitators should prioritize the goals and plan to move the process toward more ideal variations for each component.

Early Support/Expectation

This part of the implementation process tends to be controversial. No one questions the need for strong support once the implementation has been developed, defined and ready for use in the school or district. But the DESSI optimum scenario indicates that pressure must accompany the support in order to produce the highest level of implementation. "Pressure" refers not to autocratic demands from the top, but to expectations from the curriculum committee, other members of the team or department, the principal or other administrators. These expectations should be based on the understanding that appropriate use of the innovation is an important way of improving the quality of the program and no one is exempt from the responsibility of implementing it. Pressure comes in many forms and from many sources and is effective *if* it is accompanied by the support to make the necessary changes.

A major study conducted by Hall and Hord (1987) identified four major categories of support; (1) supplying materials and other arrangements, (2) providing training, (3) monitoring the implementation process and (4) consultation and reinforcement.

Supplying Materials and Arrangements

Inadequate logistical support can undermine science teaching before it begins. On the other hand, appropriate and sufficient materials provide strong encouragement for science instruction. Quality inquiry-oriented science experiments demand materials, equipment, space, and careful organization and management of these resources. An initial one time supply is not sufficient. Resources must be supplied on an ongoing basis. The ability of a district or school to provide this type of logistical support should be carefully determined well in advance of the pilot and field test. If the support is inadequate, the program must be modified in order to reduce the cost involved. An innovation that starts with marginal support is virtually doomed from the beginning.

Providing Training (Staff Development)

The preparation to use many innovations was once considered to be a one or two day orientation before school began, to provide all the information necessary to carry out the program during the entire year. There is strong evidence that this type of procedure is unsuccessful, yet it continues to be widely used. Staff development should take into account teachers' individual needs and concerns. It should be scheduled to coincide with what teachers are using in their own classroom. Finally, training should attend to adult learning factors and should be designed to help the teachers meet their own needs and develop their own personal and professional confidence.

Monitoring

The monitoring process, to a very large degree, epitomizes the support and expectation part of the model. By listening to the successes and concerns of the teacher, support and reinforcement can be communicated very directly; problems can be identified and often solved right on the spot. The presence of the principal, science supervisor, curriculum director, or the developer in the classroom of a teacher implementing the new program carries with it a strong message that someone cares about what is occurring in the classroom and expects the innovation to be used.

Consultation and Reinforcement

The central science department staff in one district called this activity "comfort and caring" (Pratt, Melle, Metzdorf and Loucks, 1980). Only so much support can be provided through formal in-service programs. Technical assistance and coaching, when conducted in the classroom, are an extremely effective part of the support structure for teachers and their implementation efforts.

Managing Implementation Based on Teacher Concerns

If the above scenario or model seems too regimented or almost manipulative, it is important that another major dimension be added. Support strategies need to be shaped according to the unique needs and concerns of the teachers involved. In the past leaders had to rely upon their intuition or the commonly held myth that ownership of an innovation would lead to a smooth, trouble free implementation process. Research from a number of sources indicates that implementation is usually not without some dissension, but it is now understood that the concerns of teachers are (1) developmental, (2) change over the duration of the implementation process and (3) can be fairly well predicted and used to guide many of the activities described above. Implementation has been the focus of a theoretical construct known as the Concerns Based Adoption Model (CBAM). The CBAM grew out of the work of Fuller (1969), who studied the concerns of preservice teachers as they moved into student teaching. She observed that their concerns evolved through a sequence beginning with concerns about themselves (*self* concerns), moving on to *task* concerns and culminating in *impact* concerns. Subsequent work by Hall, Wallace and Dossett (1973) suggested that this sequence was developmental and that it was useful in understanding the concerns of teachers as they implemented innovations in their classrooms.

The CBAM staff refined Fuller's three step model into the seven Stages of Concern about the innovation (SoC) (See Figure 6.3). To provide a quantitative measure of SoC the 35-item Stages of Concern Questionnaire was developed by Hall (1979) and others.

Figure 6.3

Stages of Concern: Typical Expressions of Concern About the Innovation

Stages of Concern	Expressions of Concern
6 Refocusing	I have better ideas about something that would work even better.
5 Collaboration	I am concerned about relating what I am doing with what other instructors are doing.
4 Consequence	How is my use affecting students?
3 Management	I seem to be spending all my time getting material ready.
2 Personal	How will using it affect me?
1 Information	I would like to know more about it.
0 Awareness	I am not concerned about it (the innovation).

The relative intensities of the sub-scale scores form a curve (known as a profile), which makes it possible to interpret the concerns of individuals or groups. Figure 6.4 presents the hypothetical patterns for a teacher who has had various amounts of experience with an innovation. Thus, a teacher who is not using the innovation would have intense Personal and Informational Concerns; a teacher just beginning to use it (an inexperienced user) would have the most intense Management Concerns; and a very experienced teacher

---

From S.M. Hord, W.L. Rutherford, L. Huling-Austin, & G.E. Hall (1987). *Taking charge of change.* Alexandria, VA: ASCD

Figure 6.4
Hypothesized Development of Stages of Concern

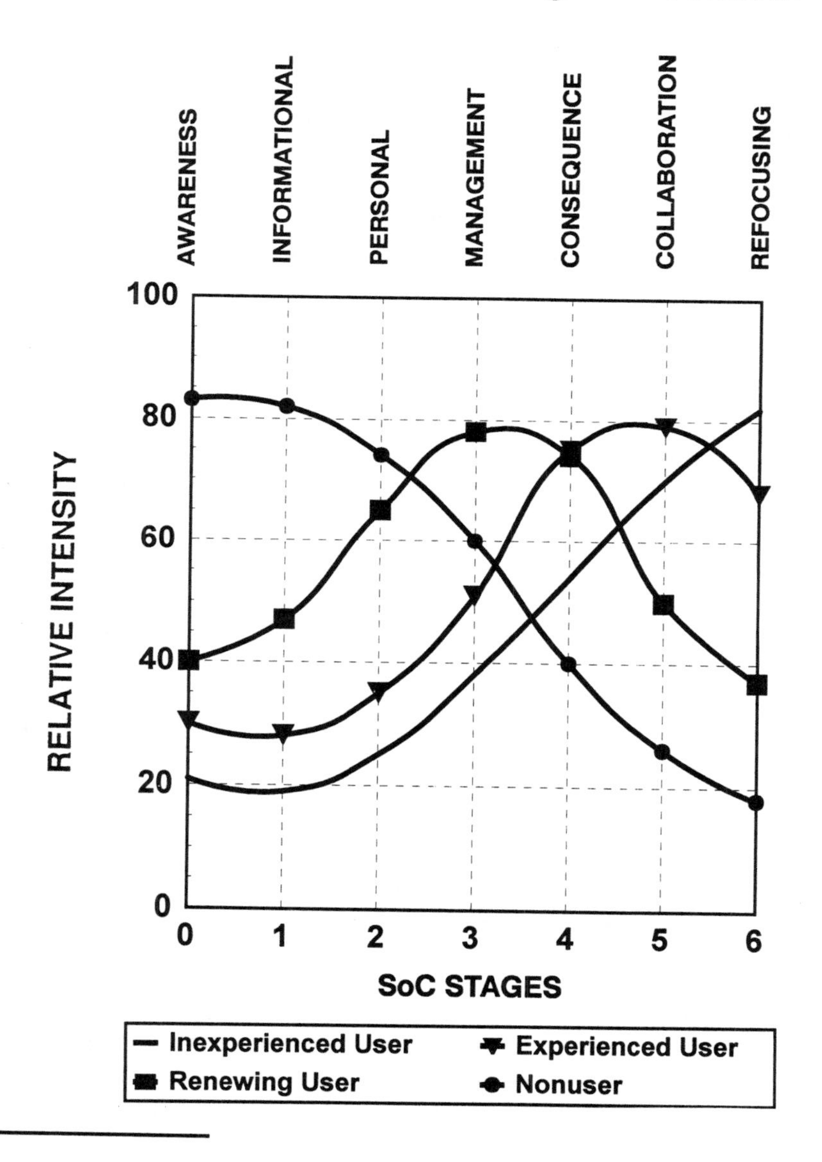

From G.E. Hall and S. Loucks, (1978). *Teacher concerns as a basis for facilitating and personalizing development*. Austin, TX: University of Texas.

would be expected to display a higher level of Consequence and Collaboration Concerns. Finally, a teacher who is redirecting his or her thinking to a new innovation or a major change in the current one will have high Refocusing Concerns. One can observe the apparent wave motion of the peak of the most intense concern moving from left to right.

The extensive research using this model verified a number of assumptions about change that are important to consider here.

1. Change is a process, not an event. The decision to develop or implement a change is not the change itself. The completion of in-service is an event, but does not immediately result in a change of behavior on the part of the participants. Change is a process that occurs over a long period of time—usually a period of several years.
2. Change is accomplished by individuals not institutions. Individuals must be the focus of attention in the intervention's design to implement a new program. Only where virtually all the individuals in a school or a school system have put the improved practice into place can it be said that the change has occurred.
3. Change is a highly personal experience. Although the developmental stages of concerns can be highly predicted, some people assimilate a new practice much more rapidly than others. Change will be most successful when geared to the diagnosed needs of individual users.

A Closer Look at the Ideal Scenario from a Concerns Perspective

Let's go back over the ideal scenario described earlier to see how it is supported and refined by a concerns based approach.

People generally have no concerns until they learn of the existence of the innovation. This lack of concern is indicated as Stage 0 Awareness (see Figure 6.3). Typically their first concern is to know if it's to be adopted or used in their school. Does it fit and meet the needs of their students and their school (Stage 1 Information). As the teachers begin to learn about the innovation they have Personal Concerns related to their ability to carry it out, the amount of time that will be required, and whether or not they will have the needed support from the principal, parents, and other teachers. If the innovation itself is well defined, dates and plans for the local school clearly spelled out, and each individual's role or expectations clearly delineated, many of these Information and Personal Concerns can be alleviated.

Nevertheless, some Personal Concerns will still persist when the first training sessions are convened. It is important to recognize this and conduct many of these sessions in small groups, with leaders who have had experience with the innovation, allowing discussion and questions concerning the new implementation by the participants.

One of the significant findings of the research is that concerns usually do not move past the Information and Personal Stages until after an individual begins to use the innovation. No amount of talking, discussing, or training will relieve the Personal and Informational Concerns people have about an innovation until they have begun to use it in their classrooms. It is virtually impossible to construct a pattern of use and a belief system about the innovation without extensive use of it.

As Personal and Information Concerns are resolved, Management Concerns (Stage 3) begin to emerge. If the materials problems have not already been resolved, Management Concerns will be paramount. If equipment is available and its distribution, storage and resupply have been arranged, many of the Management Concerns will decrease. But early in-service programs should use a direct, "how to," hands-on approach, allowing teachers an opportunity to address the Management Concerns personally and construct their own views and patterns of the innovation's use. During the monitoring, consultation, and reinforcement activities, the change facilitator can meet with individual teachers on an individual basis and address some of the Personal Concerns that still persist. Being present in a teacher's classroom also allows Management Concerns to be dealt with on the spot. Because science innovations typically are filled with new equipment and experiments, in-service sessions should be timed to anticipate and solve many of the concerns about new experiments prior to their use in the classroom. Therefore, it is important to schedule the in-service sessions throughout the year, rather than all up front in late August or early September.

Management Concerns in science curricula often take several months to resolve, but Consequence Concerns (Stage 4) emerge as teachers begin to ask questions about how to be more effective with students in their classrooms. Often these concerns must be dealt with in the classroom through coaching and personal contact. Although the agenda of the in-service sessions can take on consequence issues, experience has indicated that they are best addressed within the context of the local school and classroom.

It should be clearly understood why the interventions and strategies called for in the ideal scenario are effective. Many teachers, if left to their own resources, would approach an innovation, reach a high level of Information and Personal Concerns, and then turn back without trying it. Some will make major modifications in the innovation without ever trying it in its original form. For this reason it is important that science supervisors, principals, district administrators, and fellow teachers provide the leadership necessary to see that the innovation is used and supported. At first there will be some discontent on the part of many teachers, and some friction between teachers and administrators may occur. But as the users progress through their stages of concern and begin to develop mastery of the innovation by resolving their Management Concerns and constructing their pattern of use, the consequences of the proven innovation begin to appear in the form of greater success on the part of their students. This observable success, and enthusiasm on the part of the students, will produce a commitment to the innovation on the part of teachers. This in turn produces a long-term, stabilized use. The Concerns Based Adoption Model helps us understand why a change in behavior leads to a change in attitude on the part of teachers, and not the reverse.

A Consumer-Product or A Collegial-Constructivistism Model?

Conventional wisdom often leads some change facilitators to believe that if the decision to implement an innovation is made at the grassroots, i.e., a "bottom-up" decision that there will be good buy-in and few concerns. This is generally not true. Even self-imposed decisions to change have their stages of concerns. Concerns described in the CBAM model are probably a reflection of the progress an individual is making in constructing a new view of teaching and the behavior required to put it into practice. According to the constructivistic perspective users act upon new information by relating it to existing knowledge and personal practice, imposing their own meaning and organization on the experience. This casts the user as an active problem-solver and a constructor of his or her own knowledge, rather than as a mere passive consumer of someone else's expertise. Although information can be provided that describes the new practice, and model behaviors can be presented through demonstration teaching or videos, the new behavior can only be learned through sustained classroom practice. The SoC is a means of monitoring the growth of the newly constructed view and practice. It starts with the creation of a view of how the innovation affects the individual using it (Personal Concerns) and progresses to the development of managing (Management Concerns) it to relieve the Personal Concerns. Only then can the user construct an image and understanding how the innovation can be used to benefit the students (Consequence Concerns).

In the process of moving through this continuum of developing a pattern of use of the innovation, teachers are aided greatly by interaction with experienced teachers, staff developers, and the developer of the innovation. Fullan & Stiegelbauer (1991) stress the key role that ongoing interaction plays in making the change, noting that "the more complex the change, the greater interaction is required during the implementation." In the classroom context this collegial interaction and support often results in the mutual adaptation described earlier.

Seldom is everyone ready simultaneously to embark upon a change in curriculum, instructional techniques, or a new school structure. Typically there are those who lead, many who follow that lead, and a few who lag behind or resist. Even in a collegial, shared decision making scenario, someone is usually facilitating the change for others in the group.

The constructivistic perspective posits that it is impossible to view an innovation as one that can be "delivered" from producer to consumer and implemented with total conformity to the original design. If learning implies construction and not just receiving new behaviors, some modification or adaptation is inevitable.

Ongoing Support and Expectation

Rarely will use of the innovation continue on its own, without any future problems or concerns. For some time use of the innovation will require a greater level of effort on the part of the teachers. Management Concerns in a materials oriented science program are rarely fully resolved, and they can flare up again when equipment breaks, parts are lost, or replacement supplies do not appear on time. Futhermore, each unit has its own individual Management Concerns and issues. Soon, legitimate Consequence Concerns will arise as to whether students really are learning more than they did before. For these reasons, support such as follow-up in-services, consultation, reinforcement and evaluation studies, newsletters, equipment lists and resupply notes etc. must be provided on an ongoing basis.

In addition, the administrators and decision makers in the district must make it clear that use of the innovations is expected long after the in-services are over. Science supervisors and principals often have a tendency to turn their full attention to the next innovation and fail to provide ongoing physical, technical, and moral support to the programs underway. Many times a school district has made the decision that this is the year for science and provide hours of extra help, curriculum writing time, and in-service only to withdraw that support the following year with the assumption that the program is on it way. Good indications of support from a school district are those that begin to institutionalize the innovation, i.e., strategies that will ensure that it continues to exist and thrive long after the original supervisor, developer, and curriculum committee have left.

Apply the following check list of indicators to a program in your schools to determine if the district has a commitment to the program. The support can be expressed in a number of ways, such as:

- the presence of up-to-date curriculum guides
- instructional time allotments, particularly at the elementary level
- testing programs
- budget commitments
- ongoing administrative decisions

- newsletters
- evaluation procedures
- district policies
- in-service for new teachers
- follow-up support activities for experienced teachers
- up-to-date sources of all materials

It is very evident from the experience of the last 30 years that innovative programs that do not receive the kind of support indicated here will soon fall into disuse and be replaced by the more traditional programs in place prior to the innovation.

A Word about Decision Making

This entire chapter has been devoted to managing and supporting a change effort on the part of a school or school system. A concerns-based approach describes the developmental feelings and attitude of a group of teachers. Discussion has suggested a variety of ways to help teachers construct their new view and practices as they resolve their concerns and improve implementation of the innovation. We have been silent on how the decision was made as to what innovation to use. In no way do we mean to imply that these are made in a unilateral, single handed way by the principal, science supervisor or other administrators. We hope that just the opposite is true—that teachers have a strong, influential "say" in the decision making process. How this is handled should become clearer as we discuss leadership in Chapter 8.

References

Aikenhead, G. (1985) Teacher decision making: the case of prairie high. *Journal of Research in Science Teaching, 21,* 167-186.

Berman, P. & McLaughlin, M. (1977) *Federal Programs Supporting Educational Change.* Santa Monica, CA: Rand Corporation.

Crandall, D., et.al. (1982) *People, Policies and Practices: Examining the Chain of School Improvement.* Andover, MA: The NETWORK, Inc.

Darling-Hammond, L. (1990) Instructional policy into practice: 'the power of the bottom over the top'. *Educational Evaluation* and *Policy Analysis. 12*(3) 233-241

Darling-Hammond, L. & Berry, B. (1988) *The Evolution of Teacher Policy.* Santa Monica, CA: RAND Corporation. (quoted in Darling-Hammond, 1990.)

Dunn, W.N. & Holzner, B. (1988) *Knowledge in society: anatomy of an emergent field. Knowledge in Society,* (Spring 1988), 3-26,

Fullan, M.G. (1982) *The Meaning of Educational Change.* New York, Teachers' College Press, Columbia University.

Fullan, M.G. (1993) *Change forces: Probing the Depths of Educational Reform.* New York: The Falmer Press.

Fullan, M.G., Bennett, B., & Rolheiser-Bennett, c (1990) Linking classrooms and school improvement. *Educational Leadership, 47*(8)

Fullan, M.G., & Miles, M.B. (1992) Getting reform right: What works and what doesn't. *Phi Delta Kappan. 73*(10), 745-752.

Fullan, M.G. & Pomfret, A. (1977) Research on curriculum and instruction implementation. Review of Educational Research, 47(1),335-97.

Fullan, M.G., & Stiegelbauer,S. (1991) *The New Meaning of Educational Change.* New York: Teachers College Press.

Fuller, F. (1969) Concerns of teachers: A development conceptualization. *Review of Educational Research, 4*(2), 335-393.

Hall, G., George, A. & Rutherford, W. (1979) *Measuring Stages of Concern About the Innovation: A Manual for Use of the SoC Questionnaire.* Austin, TX: The University of Texas

Hall, G. & Hord, S. (1987) *Change in schools: Facilitating the process.* Albany, NY: State University of New York Press

Hall, G. & Loucks, S. (1978) *Innovation Configurations: Analyzing the Adoptions of Innovations.* Austin, TX: The University of Texas

Hall, G., Wallace R. & Dossett, W. (1973) *A Developmental Conceptualization of the Adoption Process Within Educational Institutions.* Austin, TX: The University of Texas

Heck, S., Stiegelbauer, S.M., Hall, G. & Loucks, S. (1981)*Measuring Innovation Configurations: Procedures and Applications.* Austin, TX: The University of Texas

Hord, S.M., & Huling-Austin, L. (1986) Effective curriculum implementation: Some promising new insights. *The Elementary School Journal, 87*(1), 97-115.

Hord, S., James, R., & Pratt, H. (1987) Managing change in the science program. In *Source Book for Science Supervisors* (3rd ed.) Washington, DC: National Science Supervisors Association

Hord, S.M., Rutherford, W.L., Hulling-Austin, L. & Hall, G,E. (1987) *Taking Charge of Change*. Arlington, VA: ASCD

Hutchinson, J. & Huberman, M. (1993) *Knowledge Dissemination and Use in Science and Mathematics Education: A literature review*. Washington, DC: National Science Foundation

Loucks, S. & Pratt, H. (1979) Effective curriculum change through a concerns-based approach to planning and staff development. *Educational Leadership*, *36*(2)

Loucks, S. & Pratt, H. (1993) Implementing a science curriculum for the middle grades: Progress, problems and prospects. In *Sourcebook for Science Supervisors* (4th ed.) Washington, DC: National Science Supervisors Association and National Science Teachers Association

Olson, J. (1980) *Innovation in the Science Curriculum*. New York: Nicols Publishing Co.

Pratt, H., Melle, M., Metzdorf, M. & Loucks, S. (1980) *The design and utilization of a concerns-based staff development program for implementing a revised science curriculum*. Paper Presented at American Educational Research Association Annual Meeting

Sarason, S.B. (1971) *The Culture of Schools the Problem of Change*. Boston: Allyn & Bacon, Inc.

7

Student Assessment and Program Evaluation

Assessment of Student Learning

Changes in education—whether in science or any other subject area—almost certainly will include testing and assessment of student learning, especially if the discussion includes politicians or members of the public. In America, tests are pervasive in the educational scene, are viewed as valid indicators of what students should learn, are thought to provide the evidence that educational improvement is needed, are expected to provide evidence of positive outcomes when reforms are put in place, and often are thought of as a powerful tool for spurring change. These assumptions about testing are widely held and rarely is consideration given to the possibility that tests as currently developed and used could be harmful to educational reform efforts.

Kevin is chair of a school science department in the third year of an extensive curriculum change endeavor. All teachers in the department are working together to develop and put in place a new program that integrates subject matter across the traditional science boundaries; utilizes constructivist notions of learning and teaching; incorporates considerably more laboratory activities than in the past; uses some new instructional approaches, especially cooperative learning; includes more attention to the applications of science knowledge to technology, societal issues, and students' concerns;

and includes some portfolio assessment. It is a big job, and the department is experiencing some frustrations in making it all happen in a manner that works.

High on Kevin's list of the difficulties they face in making the desired changes is the limited amount of time they have for their work. Their basic curriculum and instructional changes are more than time-consuming enough by themselves, but the additional time demanded by including the portfolio assessment seems to be the proverbial straw. At least that is the way some of the teachers see it. Others, however, are not willing to identify the new assessment procedures as the extra work they cannot afford; they see new forms of assessment as integral to the other changes they are making.

Lack of time is not the only source of the frustration; teachers are finding that other aspects of the assessment are problematic as well. Many of them were convinced that their new approaches to the curriculum and instruction demanded new forms of assessment, but was portfolio assessment the best way to go? Would not some other form of authentic assessment have been better? The difficulties involved in getting portfolio assessment to accomplish what they intended were more than they had expected. Students were not always sure what to put in their portfolios, why they were doing so, and how the portfolios would be used in assigning grades. The teachers were not sure how to use the portfolios most effectively for either instructional or assessment purposes. Could instruction and assessment be integrated more, how should the work in the portfolios be graded, how should students be instructed to use the portfolios, and how should the results be communicated to parents? All of this uncertainty was compounded by the fact that the State Department of Education was planning to use a form of portfolio assessment as part of the next cycle of state-wide testing, with the predictable uncertainty about what impact this would have on them and their students. The State Department of Education was out to change science education in the state, and one of the main ways they intended to accomplish this goal was with testing.

This scenario illustrates the uncertainty in assessment. Assessment is changing, but it is the area of educational change in which we are least certain how to proceed. New testing procedures are being put in place by many state level agencies, and new approaches are being experimented with at the classroom level by teachers who are not satisfied with testing and grading as we commonly know it.

At all these levels, it is safe to say that how to proceed is less clear with regard to assessment than it is with regard to curriculum and instruction.

Recent History

When the big surge of public concern about education developed in the early 1980s, testing figured prominently in the media reports. *A Nation at Risk*, among other prominent commission reports, based its pessimistic view of American education to a large extent on the results of international comparisons that showed poor American performance on various tests. Not surprisingly, many of the reforms initiated at that time were directed at increasing student scores on minimum competency tests.

As these measurement-driven reforms proceeded over the next few years, attention began to shift to the fact that the old forms of assessment were not adequate. The desire for a focus on higher level thinking and improved education—and experience with having teachers "teach to the test" when the test did not address the desired higher level educational outcomes—led to a push for new forms of assessment. Performance assessment, authentic assessment, portfolio assessment and related terms entered the conversation. The conversation has become more complex and complicated, creating a need to sort out some important distinctions.

Definitions and Connections

Although many people use the terms synonymously, some distinctions should be made between testing and assessment. Assessment is a somewhat broader term, incorporating information acquired by various means in addition to tests. Assessment involves making value judgments; the information utilized in making these judgments may come from sources that include—in addition to tests—examining student work products and observing student performance. The results of this assessment can be reported to others through such means as written statements, oral descriptions or the time-honored process of grades.

The purposes of testing vary considerably, and the tests used for the different purposes differ as well. One purpose is public accountability; many state testing programs are operated for this purpose. The intent is to provide scores to the public which indicate whether or not the schools are educating students adequately. In some cases such reporting is done by school, and the principal and teachers of each school learn how well their students performed compared to other schools. Because of the visibility of the results and the public importance attached to them, such testing sometimes is referred to as "high-stakes" testing. The nature of the tests is different than those developed for other purposes. For example, since the tests are used to make judgments about *groups* of students rather than *individuals*, the required technical characteristics of the tests differ significantly.

A second purpose of testing is selecting students for such reasons as college admission. The Scholastic Aptitude Test (SAT) and the American College Test (ACT) are the prime examples. Again, different purposes lead to different types of tests. The required technical characteristics are quite different from accountability testing. In this case, information about *individuals* is needed, and the test must meet expectations the accountability tests do not need to meet. Supposedly, these tests focus on the individual's ability and are less a reflection of a particular school curriculum than accountability tests or those given by individual teachers in a school.

A third purpose of testing is to aid students and teachers in an instructional setting. Here tests are used to determine the extent to which students have learned what was intended in a particular portion of the curriculum; teachers use the results for diagnostic purposes regarding their own teaching and to help individual students identify what additional study is needed. Tests of this type obviously should reflect the curriculum closely and be supplemented by various forms of teachers' professional judgment in such matters as assigning grades.

Our focus here is largely on the third purpose—assisting the instructional process—with a slight amount of attention to the first—accountability. Because making changes in the science instructional program is the central concern of this book, the use of test data to make judgments about the success of such changes is of interest. Whether the existing accountability tests in a given locality are appropriate for this purpose is open to question; in all likelihood they are not unless specifically designed for the purpose of evaluating a particular instructional program. It can be stated flatly that using selection tests such as the SAT and ACT for this purpose is not appropriate. Instructional testing may be helpful for program assessment purposes if designed appropriately, a matter to which we will return in the discussion of program evaluation in the second half of this chapter.

The Importance of Testing and Assessment

There are a number of reasons we must address the topics of testing and assessment when dealing with science education change. Without doing so, we jeopardize our science education reform efforts.

The *first* and most fundamental reason for addressing testing and assessment is that the "desired state" of science education advocated in the National Science Education Standards and reflected in the work of such groups as Project 2061—i.e., the reform we are pursuing—*involves a form of learning that is not measured well with the tests commonly used in our schools*. If students are to understand new goals of instruction and teachers are to be committed to teaching to them, the means of assessment must be consistent with these goals. New forms of testing are a means of communicating to others the nature of the curricular and instructional changes being sought. For teachers, the process of developing and using new classroom assessments is a powerful means of coming to understand the nature and "real world" meaning of the curricular and instructional innovations at hand. These benefits are particularly apparent when teachers collaborate with their peers and work together to pursue the various reforms.

The *second* reason is the flip side of the first one. Just as working on new forms of testing can be a means of fostering positive change, staying with current forms of assessment and failing to address testing has the potential of undermining the reforms. In their seminal publication on testing and educational reform, Resnick & Resnick (1992) address "the possibility that the very idea of using test technology as it has been developed over the past century may be inimical to the real goals of educational reform (p. 37)". Their analysis of this possibility has three components: (1) analysis of the nature of the educational goals of the reforms and what they imply for instructional activities, (2) assumptions about knowledge and competence reflected in standardized tests, and (3) the role of tests as elements in social systems. Their conclusion is very direct: "These three analyses lead us to conclude that the tests widely used today are fundamentally incompatible with the kinds of changes in educational practice needed to meet current challenges" (p 37).

Guidelines for Changing Assessments

Changing assessments is an important part of changing science education. Here are some general guidelines for doing so in an efficient and effective manner.

1. *Focus on improving assessments at the classroom level.* Although a lot of good development work is under way regarding state and district level testing of the "high stakes" variety—and this is the testing that gets most of the media attention—the assessment done by each teacher in his or her classroom is the heart of the matter. Much of the discussion in earlier chapters about improving the curriculum and instruction applies here. Teachers' concerns must be addressed. They must be involved in developing the new approaches. The process of change must be systemic; i.e., all the interrelated facets must be addressed.

   The fundamental reason for focusing on the classroom level is that the improvements we are seeking in science education constitute a fundamentally different conception of curriculum content and how it should be taught. This new conception is not compatible with traditional testing as generally practiced. If a teacher is teaching this new curriculum in the new way, a new form of assessment must go with it.

2. *Connect assessment and instruction.* Teachers tend to teach to the test, whether it is a test imposed on them or their own test. If the assessments are poor, they will tend to distort what would

otherwise be good hands-on, minds-on science teaching. If teachers are not competent in assessment, their otherwise good work on improving instruction may be undermined. Thus, there is a pressing need to provide teachers with the assistance they need in developing assessments compatible with their improved instruction.

3. *Focus on authentic assessment.* Authentic assessment—a fairly new term in the educator's lexicon—refers to assessment of tasks which are worth mastering for their own sake, not just for purposes of taking a test. They should not just focus on knowledge, but what the student *knows how to do with the knowledge.* It means assessing whether or not a student can use knowledge to solve a "real world" science problem, make an optimum decision regarding a science-related social or personal issue, design an experiment, or make an oral or written presentation of related science ideas. Such abilities are very difficult to assess with a multiple choice test; we typically must have the student perform the actual task and make judgments about the quality of the work.

   Test activities of this nature are beginning to be incorporated into many large scale tests, such as state tests. In such cases they are often referred to as performance tests, since the student is required to perform the actual task which is the goal of instruction, rather than some "test task" which is assumed to be a valid surrogate for the real task. The broader term—authentic assessment—probably better fits what a teacher generally would do in a classroom. In this case, assessment becomes an ongoing part of the instructional process, and teachers develop systematic ways of compiling their judgments about students' work on such tasks. So-called portfolio assessment is one version of this approach.

   Understanding how to do various kinds of authentic assessment is evolving rather rapidly; the person working in this arena is well advised to seek out the latest information. As of this writing, a good resource to consult is the May, 1992, issue of *Educational Leadership,* which is devoted almost entirely to this topic.

4. *Do not eliminate multiple choice tests from classrooms.* Multiple choice and other forms of objective tests measure something

different than authentic assessments. The average correlation between performance tests and traditional tests is relatively low (Shavelson & Baxter, 1992, p. 23). There is reason for both traditional and authentic assessments.

5. *Do not depend upon new tests to change the instruction.* An assumption that underlies much of the current national and state movement toward performance tests is that changing the tests will change instruction. Research in this arena, however, suggests that changing the tests is not enough (Shavelson & Baxter, 1992, p. 24). New forms of testing may be necessary, but they are not sufficient. In other words, new tests are not a substitute for the processes of curricular and instructional improvement that are the subject of the rest of this book.
6. *Provide teachers with assistance for improving assessment.* Just as assistance for teachers in making curricular and instructional improvements pays big dividends, assistance with improving assessment is a good investment. This assistance may not require an entirely new structure; assessment should be a part of instruction, and the existing support systems can be expanded to include attention to assessment along with the other support for curriculum and instruction.
7. *Foster collaboration among teachers on assessment.* Just as collaboration among teachers is a powerful tool for changing instruction, it can facilitate the process of changing assessment. The changes being sought in instruction and assessment are major; they involve changes in what many professionals currently believe about teaching, learning, and assessment. Such understandings and related shifts in values come only after a process in which teachers "reconstruct" some of their professional understandings. Dialogue with colleagues in the context of working out new instructional and assessment procedures plays a powerful role in the process of developing new understandings.
8. *Allow for the large amount of time it will take.* Authentic assessments are very time-consuming. In the case of standardized tests of this nature, that are part of state or district assessments, the cost is also high. In the case of classroom authentic assessment done by individual teachers, the time required is substantial—both to develop the procedures initially and to actually

conduct the assessments with students. Realistic expectations and allocation of resources are important to the success of the endeavor.

9. *Work hard at developing public understanding.* Fundamental changes in curriculum, instruction, and assessment cannot occur without the support of the public. Many of the changes being sought violate some public assumptions about education. In particular, parents of college-bound students tend to think that preparation for college involves being able to solve routine problems requiring certain algorithms, knowing a lot of vocabulary, and doing well on conventional standardized tests. While research indicates it is more important for students to be able to think with science content, solve novel science problems, and perform on authentic assessments, these new approaches may seem strange to the person whose schooling was different. If reforms are to succeed, the public must be educated.
10. *Maintain balance and perspective.* In the current climate there is much political rhetoric about reforming national, state, and local testing. We are faced with the problem, however, that this rhetoric "far exceeds current technological capability and ignores educational and social consequences" (Shavelson & Baxter, 1992, p.25). There is reason to think that this national and state emphasis will continue for some time. Our advice to the local leader is to be sympathetic with such reforms at those levels, cooperate fully with them when they impinge on local activities, and derive as much local benefit as possible from them, but focus local resources on helping teachers improve assessments at the classroom level.

Program Evaluation

Although assessment of student learning largely is conducted to enhance instruction, on occasion such information on student learning also may be of value in judging the worth of programs. Such judgments are at the heart of an activity commonly called program evaluation.

Sidebar 7.1

EVALUATION VERSUS RESEARCH

Evaluation and *research* are often confused. A possible reason is the similarity of techniques—such as testing and statistical analysis—often used in both of these endeavors. Conceptually, however, the two activities are quite different; they have different *purposes*. Evaluation, as defined here, has as its goal providing the information needed locally to make programmatic and management decisions. The nature of this information reflects the local circumstances in which it is needed for making specific decisions within that local context.

Research, on the other hand, is designed to provide information that is *generalizable*; that is, it applies to a much broader range of settings and time periods. While the evaluator may be preparing a report to submit to local school administrators or school boards, the researcher generally is seeking knowledge which can be published in a journal read by other persons across the nation or the world. The intent of evaluation is to provide information to be used in making very specific decisions. Research is a quest for information which will be true, under carefully specified conditions, in similar settings elsewhere. The researcher, for example, may wish to say something to the effect that instructional approach ABC works more effectively than instructional approach XYZ for teaching particular kinds of subject matter to students of particular ages, gender, academic ability and socio economic backgrounds. Occasionally, of course, data may be collected which can be used for both evaluation and research purposes. Generally, however, the motivations, constraints, and setting are such that research and evaluation activities are conducted independently.

Whenever curricular or instructional changes are made, various people begin raising questions—important and valid questions. Is this stuff any good? Should we really switch over to this new approach?

The questions are legitimate since many people are faced with a variety of decisions which must be made. Should the innovation be continued as is? Should the endeavor be modified? Should new or additional support services be provided to sustain the innovation? Should the same students and staff continue to be involved in the program? The list goes on; a seemingly endless number of decisions potentially have to be addressed. Program innovations must be evaluated to make good decisions.

The role of evaluation is illustrated by the science supervisor faced with deciding if new science curriculum XYZ, which they have been pilot testing, should be implemented across the school district, held as an option for certain schools that have a high interest in it, or totally discarded. As part of their pilot testing and related evaluation, they have been assessing student performance in the new program. Along with assessment data, information has been acquired about difficulties teachers encounter in using the new program, student interest in the program, parents' reactions to the new orientation, the cost of installing the new program, and a variety of other matters. Much of this information would not be available if a systematic process of collecting it had not been established, in some cases early in the endeavor. An evaluation process is set up with the explicit purpose of acquiring all of the various kinds of information needed to make the several decisions that are expected to arise at some point in the endeavor. A far broader collection of information is needed than simply student assessment data.

One of the distinctions commonly made by program evaluators is between *formative* and *summative* evaluations. The first of these two, formative evaluation, begins early in the process of program change. As program modifications are made, information is needed rather soon about some of its effects so decision-makers have some solid basis for deciding on potential modification in the program itself, the manner in which it is utilized, or the way in which it is supported. This information generally is not of high priority for policy setting groups such as a school board, but it is of major importance to the person in charge of day-to-day administration of a program. For this latter person, such evaluation information may be very helpful in "fine tuning" the program, e.g., deciding if a questionable unit should be modified or simply discarded, deciding if the present in-service education efforts are sufficient to show teachers how the new materials are best used, or deciding if the process of resupplying expendable materials in instructional kits is adequate.

Summative evaluation, on the other hand, comes to the fore once the new program has been "fine tuned" and in use for a period of time, say two or three years. At this point one is in a position to "sum up" the results and make a final decision about the long-term role of the new program. This summative evaluation will assist in making decisions as to whether or not a program should be continued or discontinued, expanded or enlarged with respect to the student population served, and a variety of other such decisions.

A distinction between *description* and *assessment* of the program also is helpful. Program evaluation involves making value judgments; thus the word assessment enters the picture. The measurement of the effect of an innovation on student performance is one kind of assessment. There are other kinds of assessment such as examining the impact of a new program on teachers' work loads, the degree of shift in educational objectives sought, or parental response to the new goals on which a program is based.

In addition to any assessment information that indicates whether the results of an activity are good or bad, the decision maker also has to have information about what activity actually produced this effect. Thus there is a need for descriptive information. That a new science program is said to be an "inquiry approach" does not necessarily mean students are learning according to an inquiry approach. Just that a new science program has been adopted does not necessarily mean that the instructional activities employed by teachers are being conducted in the intended manner. There is a great need for descriptive information that tells in some detail what actually is happening in the educational setting. Only with this information in hand is it possible to use assessment information to make the appropriate decisions.

The following chart illustrates a typical case in which a new science program is evaluated over the first three years of its use. The four facets of evaluation described: formative-descriptive, formative-assessment, summative-descriptive, and summative-assessment receive different emphases as indicated by double, single, and non-existent asterisks.

Evaluation Activity	Year 1	Year 2	Year 3
Formative-Descriptive	**	**	*
Formative-Assessment	*	**	*
Summative-Descriptive		*	**
Summative-Assessment		*	**

Questions to be Addressed in the Evaluation

Descriptive questions. To make decisions about modifying or adopting a program in a test phase, one must know what in fact is happening in the school setting, i.e., the extent to which the program is being conducted in the manner originally intended. There often is a considerable *difference* between actual educational practice and what was originally intended, or what is assumed to be the case. Examples of the kinds of questions which may be answered by such descriptive information include the following, as pursued in an evaluation of an activity-based elementary school science program in a medium sized school district.

- Are teachers using the program in the manner described in the teachers' guide and explained in the in-service classes, or are they still using ways of teaching they used with the old program?
- How do teachers go about obtaining the materials they use in the class each day; how long does it take them to do so?
- How much support is provided to teachers in the form of in-service education, principal or supervisor visits, and other means?

Additional examples of questions which may be answered by such descriptive information include the following, which were addressed in the evaluation of a medical careers-oriented science course in a specialized senior high school.

- What are the science topics covered in the course and what proportion of the time is devoted to each one?
- What is the nature of the various instructional activities employed and to what extent is each used?
- What type of students take the course, i.e., what are their career aspirations, academic ability, interest in future education in this area, gender, ethnicity, etc.?
- What careers are pursued by the people completing this course?

- How does the per-pupil cost of this course compare to other science courses offered in the school?

Without answers to such questions, *as reflected in actual practice*, there is considerable likelihood of erroneous assumptions being made about what is taking place. Think what this does to your decision making!

Assessment questions. In addition to descriptive information, there is a need for evaluative information about the quality and impact of what is being done. Information is needed as a basis for identifying changes which can be made to improve the educational endeavor (formative evaluation) and to decide whether the program should be adopted, discarded, or possibly modified further. Evaluative information is needed that tells *how well* it is being done and gives an indication of its success in producing student learning.

Examples of the kinds of questions which may be answered by such assessment information include the following, as pursued in the evaluation of a new applications-oriented science course in a secondary school.

- How does the amount of basic science knowledge acquired by the students compare to what they would have learned in the conventional course they would otherwise be taking?
- In addition to the basic science knowledge the students learn, how much are the students learning about societal issues and the ways in which science knowledge is used in resolving them?
- What is the impact of this course on students' interest in science and their interest in electing further science courses in the future?
- To what extent have students acquired better decision-making and problem-solving ability?
- To what extent are the teachers of the course using different instructional techniques in their *other* classes that they did not use before teaching this new course?

Answers to a substantial range of questions are relevant to making decisions about a new program.

Examples of techniques for gathering evaluation information

A wide variety of techniques are useful for collecting information about new program endeavors. Some specific examples of techniques are the following.

Records. There are a variety of records which may have been kept for other purposes and also can be of value to the evaluator in determining what events actually are occurring. For example, an elementary school implementing a new activity-based elementary school program may have a centralized system of distributing supplies and equipment, including a record keeping system and related check-out process, which leaves behind a record of all of the equipment and materials checked out by each teacher through the year. The evaluator might want to observe such records to gain an accurate perception of the extent to which the science program is being taught with the intended materials. Another useful record which may be available to the evaluator is lesson plans and related materials prepared by a given teacher who is willing to make them available. If the Level of Use instrument mentioned in a previous chapter has been used, it also provides important information of this type. Such information may provide the evaluator with indications of the extent to which a given science program is being taught—and the extent to which it is being taught in the manner originally intended by the developers of the program.

Observation. Systematic observation of instructional activities is a good means of verifying what is taking place. Such observation may utilize formal observation schedules or be based on more "free-form" observations. There are, of course, hundreds of observation schedules that have been developed for use in instructional settings, and new ones can be developed as needed based upon knowledge of the intended instruction which was to result from the innovation. For example, one observation system, known as the

Teaching Strategies Observation Differential (Anderson, et al., 1974) was developed for use with activity-based elementary school science programs as a means of recording the type of teaching strategies used by teachers of the new programs. Other such schedules can be developed as the need arises. For example, if one is introducing a program which utilizes cooperative learning techniques, a schedule could be developed for observers to use in classrooms which would provide information as to the extent such cooperative learning techniques actually are being employed. The issue may not be an evaluative one, i.e., the proficiency with which a teacher employs the particular instructional techniques. It may simply be a matter of recording information that describes the character of instruction, i.e., in this example, cooperative learning.

Questionnaires. Another source of information is self report devices, such as written questionnaires, on which personnel can respond with information about what they are doing. Again, the evaluator probably would need to develop questionnaires which fit the specific situation and which would provide various personnel, e.g., teachers and administrators, an opportunity to report the nature of the activities in which they are engaged. For example, a process of developing new teaching materials as described in Chapter 4 calls for feedback from pilot test or field test teachers via written questionnaires. Questionnaires, of course, easily can be extended beyond asking for descriptive information to asking for evaluative comments. Teachers can be asked to express some of their concerns about the new endeavor, e.g., areas of weakness that they see in materials, student responses generated, ease of teacher use of materials, etc.

Interviews. Interviews are an alternative to questionnaires for collecting self report information from the people involved. Given the time-consuming nature of this technique, it is best focused on those types of questions not readily handled with questionnaires, i.e., where greater personal attention and interaction are required to get responses. As a result, an interview likely will have more open ended questions which require considerable description, as well as interaction with the interviewer. In its more extreme form, there are few direct questions and the interviewer goes to considerable lengths to encourage the interviewee to discuss a particular topic without revealing to the interviewee a particular bias or set of values, which could occur when asking specific questions. An interviewer, for example, may initiate an interview with the following: "Please tell me something about the pains and pleasures of working with this new program." This latter form of interview can be very useful in understanding the character of the events which are occurring, as well as some of the underlying conditions which cause the activities to be as they are.

Interviews can extend into program assessment matters with ease. Interviews with students, teachers, and administrators are used to obtain valuable information about how well the program is functioning. Simple questions such as the following are often very useful in either questionnaires or interviews.

> What are the main benefits that you see resulting from this new program?
>
> What are the main difficulties that you see in being part of this new program?
>
> What recommendations would you have to offer people operating this program if they wish to improve it?

Expert judgment. An additional source of assessment information sometimes used is the judgment of outside experts. Because of their experience with a particular type of program and/or efforts for initiating and sustaining such an approach in schools, their judgments may be particularly helpful. Sometimes this assistance is acquired through a formal review process that may involve a team of outsiders, possibly as part of a routine periodic review that may be used in a given school district. In other cases, the outsider may be brought in as an individual, simply to review a particular innovative program and offer suggestions in that context. In either case such outsiders often have a valuable perspective to offer and can identify important matters which are not apparent to people who must address the same issues on a day to day basis.

Student assessment scores. Test scores often are the "bottom line" policy makers are interested in using to make their judgments. Given this attention to test scores, it is absolutely essential that careful consideration be given to obtaining measures with as full a range as possible of the objectives from the program. Thus, much of what was said earlier in the section on student assessment—particularly with respect to authentic assessment—is of crucial importance in this context. If, for example, one of the objectives is to give students an understanding of science related societal issues or an understanding of the personal applications of science knowledge, assessments must measure adequately these objectives. A test designed for the conventional program is not an appropriate basis for judging the success of the applications-oriented course. A great deal of extra effort may be required to obtain the appropriate assessment. This extra effort is crucial if the assessment is to reflect the educational objectives of the program.

If comparing two curricula—such as an applications oriented course and a more conventional one—the assessment must address all of the objectives of *both* of the courses. Once information is obtained on the extent to which the two courses have enabled students to meet the applicable objectives, it is possible to make some value judgments about how well the two courses are functioning. But it is crucial to realize that it is not simply a matter of seeing which of two courses produces the highest student scores on some test judged to be a good compromise for the two courses. If, as often happens, both courses are found to do an acceptable job of reaching their intended objectives, it is a matter of making some value judgments about which objectives you want to have a program to pursue. In terms of the quantity, quality and appropriate dimensions discussed in an earlier chapter, it is a value judgment about what objectives are appropriate.

Who conducts the evaluation?

Although a comprehensive evaluation endeavor may be conducted entirely by local district personnel, it is common for outside evaluation specialists to be involved as well. Formative evaluation typically is managed by local district personnel although outsiders sometimes play a role in designing it. In fact, the most important role an outside evaluation specialist can play in all aspects of an evaluation is in its design early on. An outside specialist with expertise in both evaluation and science education would be particularly valuable.

In districts large enough to have their own evaluation specialist or even evaluation staff, they routinely handle this type of work. Given the nature of science innovations under consideration here, however, it is unrealistic to expect that such evaluation specialists also will have the expertise in science, curriculum and instruction required to do this job. The significant involvement of a science specialist within the district is needed if the evaluation is to be what it should be.

In the case of the summative evaluation, there may be reason for more full involvement of outside evaluation experts, particularly with externally-funded endeavors when the funding agency may wish to have this work done by an outsider with a certain kind of credibility and independence. If a local education agency has the evaluation expertise on its staff, of course, there is no reason they cannot do this type of work if this approach is acceptable to the funding agency.

Audience for the evaluation

The key audience for the formative evaluation is made up of the various personnel responsible for developing and managing the program under consideration, including teachers, principals, science department chairpersons, science supervisors, curriculum directors, and other administrators. The audience for the summative evaluation includes these same participants and administrators involved in operational decisions, but in addition includes policy makers, e.g., a school board, which must make decisions about program continuation, termination, or expansion.

Means of reporting

The means of reporting this information to decision makers is an important consideration. Many means are possible although standard written and oral reports are the most common. There also may be a significant place for illustrative examples in the form of video tapes of brief segments of various instructional activity which characterize what in fact occurs.

The process does not end with simply presenting the information. In the case of formative evaluation, the decision makers—both teachers and administrators—need to have an opportunity to "digest it" and assess what changes they will make based on this information. There is a need for discussion of the results and deliberation over the steps which could be taken to act upon the new information. Often the process is assisted by small group discussion, focused on specific facets of the report. Within this setting, careful consideration can be given to the many alternatives available, and the various people can gain the insights of their co-workers for each of the individual situations.

Because of the different audience for summative evaluation results, the reporting of results may take a different form. When there is a need to make a formal presentation of the results, it may well be to a fairly large group. As a result there is a place not only for carefully prepared written reports, but slide tape shows, effective transparencies for overhead projectors, and other such devices which will aid in presenting the results of the evaluation of the innovative program.

Some Concluding Observations

At some point in the process of putting a new educational endeavor in place, decisions must be made as to whether or not the new program lives up to prior expectations and is worthy of continuation. Should it be abandoned, possibly in favor of something it replaced, or should it be retained on a more or less permanent basis? If the new program was set up originally on a trial basis for a limited clientele, the resulting decision may be one of expansion to potential participants within a given constituency. A crossroads has been reached; results must be summed up and an assessment made which can serve as a basis for these crucial decisions.

One of the observations of persons familiar with evaluation in educational settings is that the results of an evaluation process often are not used. Perhaps the results of the evaluation are not available at the time of making a decision. Perhaps political considerations, or an inadequately done evaluation cause the evaluation to be of little consequence. As a result of this phenomenon, the person responsible for evaluation should design the activity from the very beginning in a manner that takes account of such considerations. Careful planning must insure that the results of the evaluation will be fed into the decision-making process at the necessary points in time. Formative evaluation results should be available early in the process, and on a continuing basis, so that decision makers have the information when needed. Summative evaluation results must be available at the time final decisions will be made. These considerations necessitate that the evaluator be fully aware of the projected time schedule and plan the evaluation to meet these deadlines without fail. Plans for the evaluation work must be initiated at the very beginning of the project.

Similarly, the evaluator must be aware of many political considerations that may enter the picture. To whatever extent these considerations are a factor in the evaluation process, they must be taken into account. Every effort must be made to anticipate what factors will influence a decision. If there are any data that could be collected relative to these factors, their collection should be built into the evaluation design. It is important that the evaluation be a quality endeavor with results both valid and credible.

In summary, evaluation plays an important role in the process of introducing improvements into science programs. The evaluation endeavor may fall short of the mark and have little impact unless it is given high priority and conducted in the best possible manner. Thus, it is essential that when formal evaluation procedures are employed, all the various kinds of technical expertise are acquired as needed and sufficient resources are devoted to the job to accomplish the intended results. If evaluation cannot be done well, there is probably little point in doing it. On the other hand, if done well, it can be of great assistance in achieving better decision making and better science programs.

References

Anderson, R.D., J.A. Struthers & H. H. James. (1974) The teaching strategies observation differential. In Gene Stanford and Albert Roark, *Human Interaction in Education*, pp. 274-280. Boston: Allyn and Bacon, Inc.

Resnick, L.B. & D.P. Resnick. (1992) Assessing the thinking curriculum: New tools for educational reform. In B.R. Gifford & M.C. O'Connor (Eds.), *Changing Assessments: Alternative Views of Aptitude, Achievement and Instruction* (pp 37-75). Boston: Kluwer Academic Publishers.

Shavelson, R.J. & G.P. Baxter. (1992). What we've learned about assessing hands-on science. *Educational Leadership, Vol* 20-25.

8

Facilitating Science Improvement: The Leadership to Get the Job Done

In some respects this chapter is a review of the entire book. In the preceding chapters we laid out what needs to be done. We described ways of conducting a needs assessment, setting goals, developing, selecting or adapting materials, improving instruction, implementing a program, maintaining the innovation and evaluating the effect of the change. The steps and information are there, but one very essential ingredient is missing, who makes it happen? These steps do not spontaneously occur, nor does even a well informed professional and ambitious staff carry out these functions without the important ingredient of leadership.

In this chapter we will describe the possible leaders and how they function at various levels within the state, local school district, individual school level and universities. We will describe what the leaders do and what differentiates a leader from a manager.

The reader should be reminded that the perspective from which this book is written is that an improvement in a program is a combination of change in materials and a change in people's behavior. The focus of this chapter will be largely on what leaders do to bring about a change in the behavior of the people with whom they work. Our perspective is that behavior changes by a group of people don't occur spontaneously. Somebody facilitates, supports, encourages

Sidebar 8.1

Leadership Is:

Making things happen, or not happen;

Getting others to do what they ought to do, and like it;

Making people think things are possible that they didn't think were possible;

Getting people to be better than they think they are or can be;

Inspiring hope and confidence in others to accomplish purposes they think are impossible;

Perceiving what is needed and right, and knowing how to mobilize people and resources to accomplish these goals;

Creating options and opportunities, clarifying problems and choices, building morale and coalitions, providing a vision and possibilities of something better than currently exists; and

Empowering and liberating people to become leaders in their own right.

and sometimes insists that the change in behavior occur. As pointed out in Chapter 6, there is a very easy trap into which many inexperienced leaders and managers fall. If the innovation looks appealing, e.g., if the new materials work well in a few classrooms and are supported by a committee of teachers, it will be easy to successfully implement the innovation in all classrooms. The research data from a number of sources (Berman and McLaughlin, 1977; Crandall, 1982) are all very clear that much effort and time on the part of a leader or

---

From Bybee, R. (1993) *Reforming Science Education: Social Perspectives and Personal Reflections.* New York: Teachers College Press.

group of leaders is necessary to bring about a significant improvement in the program.

The focus of this book is on change at the local school level, and as such must attend to leadership at both the district and school level. The role of the principal, particularly at the elementary level, in leading improvement efforts has been well established. At the secondary level the department chair often eclipses or supplements the role of the principal in improvement efforts. The leadership also comes from personnel in the district office, from external facilitators, such as university personnel, specialists from intermediate agencies, or other consultants. Leadership often is exercised from the state level although this leadership, along with most other leadership from outside the school system requires definite linkages with leadership inside the system. There will be more discussion on the role of these various leaders later in the chapter.

What do the Leaders/Facilitators Do?

A number of studies (Cunningham 1985; Hord and Hall, 1982; Hord, Hall and Zigarmi, 1980) have documented the actions taken to support curriculum implementation or program improvement activities. These studies indicate that six distinct categories of interventions are important for supporting change. They are:

1. Creating a vision of the improvement.
2. Providing necessary logistical materials and other supportive arrangements.
3. Providing adequate staff development.
4. Monitoring.
5. Consulting and reinforcing.
6. Developing external communication.

The task of the leader is developing a vision and communicating it to all those involved. The vision should include what the change will be, what it looks like, its value, its importance, and most of all, what will be different when the improvement has been made. The degree to which this activity or characteristic is present is probably the

Sidebar 8.2

Twelve "dimensions of supervisory practice" emerged from ASCD-sponsored research:

1. Communication, ensuring open and clear communication among individuals and groups throughout the organization.
2. Staff development, developing and facilitating meaningful opportunities for professional growth.
3. Instructional program, supporting and coordinating efforts to improve the instructional program.
4. Planning and change, initiating and implementing collaboratively developed strategies for continuous improvement.
5. Motivating and organizing, helping people to develop a shared vision and achieve collective aims.
6. Observation and conferencing, providing feedback to teachers based on classroom observation.
7. Curriculum, coordinating and integrating the process of curriculum development and implementation.
8. Problem solving and decision making, using a variety of strategies to clarify and analyze problems and to make decisions.
9. Service to teachers, providing materials, resources and assistance to support teaching and learning.
10. Personal development, recognizing and reflecting upon one's personal and professional beliefs, abilities and actions.
11. Community relations, establishing and maintaining open and productive relations between the school and its community.
12. Research and program evaluation, encouraging experimentation and assessing outcomes.

These dimensions were identified from an exhaustive search of the literature and then confirmed and ranked in the above order through a survey of over 1000 teachers and supervisors. Is there a match between the ASCD list and the six categories?

From E. Pajak (1990) Dimensions of supervision. *Educational Leadership,* 48(1) 78-80.

difference between just managing an improvement effort and leading it. Cronin (1984) emphasizes the importance of goal-centeredness and notes that many managers are often preoccupied with skills and skill development at the expense of a strong commitment to defining purpose and clarifying goals. The old axiom that if "you don't know where you are going, any road will take you there" aptly describes the importance of defining and communicating a vision. Cronin goes on to point out that a repertoire of leadership skills that do not follow clearly defined goals are of little use and maybe even dangerous.

The second activity of providing psychological and logistical support needs no explanation for most science educators. Providing the necessary budget to buy materials, textbooks, and equipment is obviously a prerequisite for many innovative science activities and curriculum. Facilities in which science is taught are also critical for most programs. In addition to these more concrete aspects, there are other equally important planning decision activities that occur within this category. Who schedules the laboratory or the science room at the elementary school if more than one person is going to use it? Who is responsible for maintaining the organization of the store room and reordering new materials? Who is assigned to teach the new course? All these arrangements often have a very significant impact on the ability of a staff to carry out an innovation.

The reader of this book should not have to be convinced of the importance of staff development. Chapter 6 points out that staff development needs to:

1. Fit the specific innovation or improvement effort. It needs to teach the specific behaviors and skills called for by the innovation.
2. Be continuous, not just up front. The model of staff development in Chapter 6 advocates early support and continuous ongoing in-service training. The initial training should be phased in over a period of time so that it fits the developing concerns of individual teachers. Follow-up in-service later on needs to have problem solving dimensions that can clarify innovation misconceptions about the innovation and meet the users' higher level stages of concern that emerge months and often years into the program.

Monitoring the program is critical. This serves at least three major functions in making the improvement effort successful:

1. *Coaching.* The presence of the leaders/facilitators in the classroom of the teachers implementing the innovation communicates the importance of the innovation to the leader. Although the very presence of the individual is a powerful means of communicating, the interaction between the facilitator and teacher can often directly communicate the importance of the innovation, the philosophy or vision behind the innovation and the technical know-how that is so important to the user.
2. *Obtaining feedback.* Visits with classroom teachers and observation of their teaching provides an opportunity to listen to teachers' needs, concerns, and ideas, and obtain direct feedback about the progress teachers are making toward carrying out the new behaviors. This feedback often is in the form of direct conversation with teachers where they express their concerns or awkwardness about some aspect of the program. Consultation and reinforcement go hand-in-hand with monitoring and often are carried out simultaneously. Although it is easy to find the problems in implementing the innovation during a monitoring session, it is important to take the opportunity to provide positive reinforcement. Coaching individual teachers is the art of finding what is working well and reinforcing it.
3. *Supporting.* While you are listening you will hear needs related to implementing the innovation. Grab that opportunity to be of direct help! Maybe some supplies are missing, maybe the directions in the guide are not clear; or maybe the reason for a particular strategy is not spelled out. If you cannot answer the question on the spot promise to get back to them in 24–48 hours.

Sidebar 8.3

In *Passion for Excellence*, Peters and Austin (1985) describe the purpose of their favorite strategy, management by walking about (MBWA).

> To begin with, as the effective leader wanders/coaches/develops/engenders small wins, a lot is going on—at least three major activities, usually all at once. They are (1) listening, (2) teaching and (3) facilitating. Listening is the "being in touch" part, getting it firsthand and undistorted—from suppliers, customers and your own people. Varieties of listening was the subject of Chapter 2, 'MBWA: The Technology of the Obvious.' The very act of listening suggests a form of caring. MBWA is also a 'teaching' (and 'coaching') act. Values simply must be transmitted face-to-face. The questioning routines, order of visits and a host of other variables add up willy-nilly to the teaching of values. Finally, the wanderer can also be of direct help! The role of the leader as servant, facilitator, protector from bureaucracy (and bureaucrats) is the third prime MBWA objective.

Classroom teachers are not the only clients of a successful leader/facilitator; communication also is needed with external audiences such as parents, school board members, other administrators, external facilitators, material suppliers, and others in order to gain support for the innovation. Community support, official board of education action, the superintendent's concurrence, necessary budget arrangements, etc., are important to the short- and long-term success and support of the improvement effort.

Who are the Leaders/Facilitators and What Do They Do?

Who carries out the various functions and activities described in the above section? The strongest answer is: *Everyone*. If science education reform is to succeed, Bybee (1993) believes that one key element is *distributed* leadership, wherein all individuals must contribute to the improvement effort as "leaders." By defining leadership as an individual's ability to work with others to improve science teaching and learning and to accomplish the goal of scientific literacy, virtually everyone in the science education community is included.

A number of studies have documented the forms of assistance provided by principals and central office administrators. But leadership can come from a variety of levels, depending on the organization and size of school system and state involved. At the high school level, and to some extent at the middle school level, leadership for program improvement is likely not to come from the principal, but from a department chair or an influential teacher, whom most department members view as a leader with strong influence on the major direction the department takes. A study from the Center for Research on the Context of Secondary School Teaching (McLaughlin and Talbert, 1993) indicated that in many comprehensive high schools teachers consider their department their primary workplace, not the school as a whole. With this in mind, the following studies should be read as a review of the *functions* needed to bring about improvement or reform, regardless of the administrative structure of the district or local school. In many respects the department chair could be substituted for the principal in the following report. If district level administrators are not involved or do not exist, the roles outlined in that section must be fulfilled at the school level. The comprehensive, systemic nature of the successful change process cannot be overlooked just because the "system" is small.

In a study of Dissemination Efforts Supporting School Improvement (DESSI) reported by Bauchner and Loucks (1982), principals reported types of assistance they provided to support a specific practice or improvement effort. Table 8.1 lists these forms of assistance in a descending order of frequency mentioned by a sample of 108 principals involved with implementing an innovative practice in their buildings.

Table 8.1

Forms of assistance provided by principals (in descending order of frequency). The categories of support described earlier in the chapter are indicated in parentheses after each form of assistance. (Note: Although the principal was the target of this study, the researchers also noted the frequent presence of a second supporting leader who could be an assistant principal or department chair.)

1. Communicate with staff. (#1-creating a vision and#5 consultation and reinforcement)
2. Plan, schedule, organize. (#2-logistical support)
3. Provide resources. (#2-logistical support)
4. Leverage staff. (#2-logistical support)
5. Provide support. (#2-logistical support)
6. Attend training sessions and meetings. (#4-monitoring)
7. Observe the program in classrooms. (#4-monitoring)
8. Handle paperwork. (#2-logistical support and #6-external communication)
9. Arrange training. (#3-in-service training)
10. Communicate with external facilitators. (#6-external communication)
11. Audit program. (#4-monitoring)
12. Make recommendations. (#1-creating a vision and #5-consultation and reinforcement)

The research results are abundantly clear; the site of effective improvement in a school district is the local school led by a skillful principal. The individual teacher, district level coordinator, and/or other administrators may play a key role but the success of an improvement effort depends on skillful interventions of the individual school principal.

Although an innovative activity or school improvement project may originate at the state or district level, the principal must support it all the way for the activity to succeed. The pressure, support, and follow up (outlined in Chapter 6) for successful implementation are very difficult to effect at anything other than the local school level. Skilled principals reinforce a vision of the innovative program being implemented regardless of whether it be a local school activity, district-initiated, or one that comes from the state. They will find ways of getting their teachers to the in-service courses, sometimes even filling in for a teacher in the classroom so he or she can be gone for portions of the day. They make trips to neighborhood schools to borrow equipment, microscopes and teacher's guides when they are not available in the local school. They visit classrooms, discuss lessons with teachers, ask questions that reflect both knowledge of the curriculum and understanding of what it takes to implement it. They let parents know what is going on in their school, support the actions of their teachers with the PTA and local advisory committees, and communicate through newsletters, personal conversations and responses to phone calls.

In a parallel study also conducted by the DESSI project, Loucks and Cox (1982) reported a similar list of activities conducted by district level personnel, mostly curriculum coordinators. Table 8.2 indicates how district level facilitators spent their time in assistance activities involved in local school improvement efforts.

Table 8.2

Assistance activities of district facilitators. Again, the intervention categories are indicated in parentheses after each form of assistance.

I. Assistance in deciding on new practice. (#1-creating a vision and #2-logistical support)
   A. Seeking commitment from school administrators.
   B. Seeking commitment from teachers.
   C. Seeking support from the local school board.
   D. Preparing a case for the decision to adopt.
   E. Assessing needs.
   F. Building support among school personnel.
   G. Making library and computer searches for materials.

II. Assist in preparing for adoption. (#2-logistical support; #3-in-service training)
   A. Arranging training.
   B. Training users.
   C. Providing detailed information.
   D. Securing materials and other required resources.
   E. Working with administrators.
   F. Working with site contact person.
   G. Allocating financial resources.
   H. Maintaining support among school personnel.

III. Assist in implementing. (#2-logistical support)
   A. Planning implementation schedules.
   B. Providing technical assistance or follow-up training.
   C. Assisting teachers in working out procedural details.
   D. "Putting out fires."
   E. Maintaining support among school personnel.

IV. Follow-up activities. (#4-monitoring; #2-logistical support)
    A. Collecting impact data.
    B. Analyzing impact data.
    C. Assisting local site in evaluating a practice.
    D. Developing plan to support continuation of new practice.
    E. Identifying additional new users at site.

From the Table 8.2 it can be seen that district facilitators were active in every phase of deciding, preparing, implementing and follow-up. During the implementation phase, the district facilitators were somewhat less active than they were in the earlier phases. At this phase the local school administrator or principals were probably the most active facilitators, but again in the follow-up phase the district level facilitators become more involved, especially in planning for continuation of the practice. What does this table tell us? It describes a group of individuals who "get their hands dirty" working in the school with teachers and administrators. They are the cheerleaders, building and maintaining commitment and spirit. They bring the new practices to the schools and teachers and they act as troubleshooters and resource personnel.

One of the most interesting findings in this study was the impact of the district level facilitators on two outcomes; (1) the change in classroom teaching, and (2) the number of benefits that teachers attributed to the new practice. It was discovered that the amount of assistance the district facilitator provided to schools was significantly related to the teachers' change in practice and the benefits teachers perceived, noticed or observed from the practice. It appears that the more time the district facilitator spends training and arranging for training and the more energy he or she puts into working with administrators and the local teachers, the more sophisticated, skillful, and successful the teacher-users feel about the practice itself.

The leaders from large school districts where there are central office personnel often assigned to individual subject areas such as science, mathematics, etc. will wonder what the relationship between the principal and district level coordinator should be. Although there is very little research to answer this question, it is our experience that the six categories of support outlined at the beginning of this chapter must be in place at the local school level. Certainly the local principal or department chair does not always create and single handedly perform each of these functions, but they are ultimately responsible to see that they occur. The central office coordinators find themselves in a consultation or assisting role. The district science coordinator may indeed create and write a vision for an improved science program, but it is his or her responsibility to assist the individual principal in understanding and communicating that vision to teachers. The in-service training may be conducted by an outside consultant or by the district coordinator, but if support is inadequate, or arrangements are not made by the principal so that teachers can attend, the in-service activities will not benefit the local teachers.

Some work has been done to specifically identify the role of the science supervisor at the local district level. Madrazo and Motz (1983) reported on a national survey in which teachers and administrators were asked to rank the roles of science supervisor in the order that would fit their needs. In order of preference these roles were:

1. Instruction
2. Curriculum
3. Staff development
4. Implementation
5. Management
6. Assessment
7. Assignment, transfer and equalization of teacher load.

This list is very similar to the more extensive list of activities from the DESSI study.

Developing the Capacity of a School or System to Change

Effective schools or school systems are often described as those that have the capacity to change or bring about ongoing, continuous improvement. One of the results of effective leadership is the development of the capacity of an organization to undergo continuous reform.

Fullan (1985) has compiled a set of eight organizational variables and four process variables that he attributes to effective schools that have developed a capacity to improve, a step beyond the ability to implement a new program.

Organizational Variables

1. Instructionally focused leadership at the school level.
2. District support.
3. Emphasis on curriculum and instruction.
4. Clear goals and high expectations for students.
5. A system for monitoring performance and achievement.
6. Ongoing staff development.
7. Parental involvement and support.
8. Orderly and secure climate.

Process Variables

1. A feel for the improvement process on the part of leadership.
2. A guiding value system.
3. Intense interaction and communication.
4. Collaborative planning and implementation.

Senge (1990) has pointed out the importance of building a learning organization. Such an organization must first have the capacity of adaptive learning or the ability to cope with the environment; but more importantly, it must move on to generative learning, which requires the ability to create new outcomes and new ways of looking at the world. Leaders in a learning organization are committed to long term systemic perspectives and are less likely to be viewed as charismatic, short-term problem-solving heroes—our traditional image of leadership. Senge describes leaders in learning organizations as designers, teachers, and stewards who are intensely concerned with establishing processes that ensure that purposes, policies, and strategies of the organization are continually improved.

How Leaders get the Job Done

There is more to accomplishing an improvement project than simply carrying out the various functions and activities listed in the section above. Effective leaders go about the job differently than their less-effective colleagues. Hidden behind each of the items in the list from the last section are the styles, intuition, experiences, knowledge and enthusiasm that a leader brings to the particular task or activity. Many writers have attempted to describe the art and style of leadership and probably nobody has accomplished it more eloquently than Peters and Waterman (1982) in the popular book *In Search of Excellence*. Through a series of well documented and vividly portrayed examples, Peters and Waterman outline the holistic and integrated approach that effective leaders bring to their task. They state that an

> Effective leader must be the master of two ends of the spectrum: ideas at the highest level of abstraction and actions at the most mundane level of detail... It seems the only way to instill enthusiasm is through scores of daily events with the value shaping manager becoming an implementor par excellence. In this role the leader is a bug for detail and directly instills values through deeds rather than words: no opportunity is too small, so it is at once attention to ideas and attention to detail.

In a study of principals Hall and Rutherford, (1983) described three facilitator styles and their effectiveness in carrying out the implementation of an innovation. Based upon a study of a group of elementary principals, Hall identified and defined three different facilitators' styles. They are as follows:

Initiators

> Initiators hold clear, decisive, long range goals for their schools that transcend but include implementation of curriculum innovations. They have a well-defined vision of what their school should be like and what teachers, parents, students, and the principal should be doing to help the school move in that direction. Initiators push; they have strong expectations for students, teachers, and themselves and they push to see that they are all moving in goal oriented directions. They convey and monitor these high expectations through frequent contact with teachers and clear explication of how the school is to operate and how teachers are to teach.

Managers

Managers demonstrate responsive behaviors to situations or people, and they also initiate action in support of a change process. Variations in their behavior seem to be linked to their rapport with teachers and the central office staff, as well as how well they understand the purpose of a particular innovation. They are efficient and provide basic support to teachers. They keep teachers informed about decisions and are sensitive to teachers' needs. They protect their teachers from excessive demands..., but once they understand that outsiders want something to happen in their school they become involved with their teachers in making it happen. They do not typically initiate attempts to move beyond the basis of what is imposed.

Responders

They are concerned about how others will perceive decisions and the direction the school is taking. They therefore tend to delay decisions to get as much input as possible and to be sure that everyone has had a chance to express their feelings. They believe their primary role is to maintain a smoothly running school by focusing on traditional administrative tasks, keeping teachers content and treating students well. Another characteristic of responders is the tendency toward making decisions based on immediate circumstances rather than on long range instructional or schools goals.

When the successes of a number of principals in implementing innovative programs were compared, the principals with the initiator style were the most successful, with the manager style the next most successful, and the responder style a poor third. This study, together with the many anecdotal descriptions from Peters and Waterman and other authors, seems to validate the notion that how leaders (not just principals) go about carrying out their functions does make a difference. Robert Evans (1993) has identified five biases for action that foster innovation.

Clarity and Focus. Attention to goals helps provide a clear and compelling vision. Innovations and reform efforts need not be narrow and limited, but a unifying focus is necessary to communicate the priorities and avoid fragmentation and distracting diversions.

Participation—not Paralysis. Collaboration is vital to reform efforts, but the leader must be ready to nudge the group on to the next step or slow the pace if too many members are being left behind, and group cohesiveness is suffering.

Communication. The larger the innovation, the greater the need for communication. An effective leader will keep the goals and plans in front of the group at all times and at the same time be an available and receptive listener. At times the communication needs to be formalized through advisory committees or evaluation and monitoring instruments, but it is always characterized by openness and candor.

Recognition. In the early stages of an innovation when personal concerns are the highest, teachers and others need confirmation that they are on the right course and rewards for the extra effort required to make the necessary changes. Leaders recognize innovations require more effort and different behaviors by all involved and that recognition is needed to sustain the effort.

Confrontation. In spite of the positive moves made by leaders outlined above, there will still be resistance that must be confronted. To avoid it is to lose credibility and establish a norm of avoidance that will deter the organization from moving to higher levels of learning capacity.

References

Baucher, J. & Loucks, S. (1982) Building administrators and their role in the improvement of practice. Paper presented at the annual meeting of the American Educational Research Association.

Berman, P. & McLaughlin, M. (1977) *Federal Programs Supporting Educational Change.* Santa Monica, CA: Rand Corporation.

Burns, J.M. (1978) *Leadership.* New York: Harper and Row Publishers.

Bybee, R. (1993) *Reforming Science Education: Social Perceptions and Personal Reflections.* New York: Teachers College Press.

Covey, S. (1990) *Principle Centered Leadership.* New York: Summit Books.

Cox, P.& Havelock, R. (1982) External facilitators and their role in the improvement of practice. Paper presented at the annual meeting of the American Educational Research Association.

Crandall, D., et.al. (1982) *People, Policies and Practices: Examining the Chain of School Improvement.* Andover, MA: The NETWORK, Inc.

Cronin, T. (1984) Thinking and learning about leadership. *Presidential Studies Quarterly,* Winter 1984, 22, 33-34.

Cunningham, L. (1985) Leaders and leadership: 1985 and beyond. *Phi Delta Kappan, 67*(1) 17-20.

Evans, R. (1993) The human face of reform. *Educational Leadership, 51*(1).

Fullan, M. (1985) Change process and strategies at the local level. *The Elementary School Journal, 85*(3) 391-421.

Hall, G. & Rutherford, W. (1983) Three Change Facilitator Styles: How Principals Affect Improvement Efforts. Austin, TX: The University of Texas.

Hord, S., Hall, G. & Zigarmi, P. (1980) *Anatomy of Incident and Tactic Intervention,* Austin, TX: The University of Texas.

Hord, S. and Hall, G. (1982) *Procedures for Quantitative Analysis of Change Facilitator Interventions.* Austin, TX: The University of Texas.

Loucks, S. & Cox, P. (1982) School district personnel: A crucial role in school improvement efforts. Paper presented at the annual meeting of the American Educational Research Association.

Madrazo, G. & Motz L. (1983) Do we link school science to science supervisors as a resource? In NSTA Yearbook; *Science teaching: A profession speaks* (pp.76-78) Washington, DC: NSTA.

Pajak, E. (1990) Dimensions of supervision, *Educational Leadership, 48*(1) 78-80.

Peters, T. (1987) *Thriving on Chaos.* New York: Harper and Row

Peters, T. & Austen, N. (1985) *A Passion for Excellence.* New York: Random House.

Peters, T. & Waterman, R. (1982) *In Search of Excellence.* New York: Random House.

Pratt, H. (1984) Science leadership at the local level: The bottom line. In *NSTA yearbook: Redesigning science and technology leadership.* Washington DC: National Science Teachers Association.

Senge, P. (1990) *The Fifth Discipline: The Act and Practice of the Learning Organization.* New York: Doubleday.

Sergiovanni, T. (1991) *Value-added Leadership: How to Get Extraordinary Performance in Schools.* New York: Harcourt Brace Jovanovich.

9

Barriers and Pitfalls

In the earlier parts of this book we presented chapters on the research and the practical ways of doing needs assessments, developing or adopting curriculum, improving instruction, and implementing or institutionalizing the change. To the extent possible, each chapter was based upon the positive results of both research and successful experiences. To a large extent this chapter will be the opposite. In over seventy years of collective experience, the two authors have observed a number of barriers and pitfalls, some of them rather common, we feel should be pointed out as negative examples. We hope that through the use of some counter examples we can better define and reinforce the positive procedures advocated earlier. Before looking at some specific pitfalls, it will be useful to examine some general barriers to improvement.

Barriers to Improvement Efforts

Facilitators of change need to recognize the following potential barriers and find ways to avoid their negative influences.

Inertia. Making change is difficult; basically, people do not like significant change. It is much easier to follow past patterns than to develop new ones. While the current patterns may not be perfect, experience has already shown them to be better than many of the alternatives. Thus, the general tendency of most people is to shift from their current routines only when there is considerable indication that it is worth the many risks involved.

Past history. Previous experiences taught most of us to be cautious about change. This is particularly true where one or more innovations has been attempted in a school setting and resulted in failure. Such experiences make it especially difficult for the next promoter of change. A common reaction to suggestions of innovations in such a setting is usually reluctance and even overt resistance.

Teacher concerns. Teachers' fears and reticence are not surprising; in fact, they are to be expected. The wise innovator recognizes their existence and their potential as barriers to innovation and takes appropriate steps to address them. In this regard, the CBAM (Concerns Based Adoption Model), outlined in Chapter 6, is of particular help. Teachers' concerns vary over the life of an innovation, and appropriate steps must be taken to address the relevant ones at each point, to avoid undue influence from major concerns.

Supplies and equipment. A very appropriate concern among science teachers relates to the availability of supplies and materials. Given sufficient resources and administrative acumen, this barrier is not particularly difficult to overcome. On the other hand, the lack of these "givens" is a major barrier to science education change.

Lack of assistance. Another potential barrier to implementation is a lack of sufficient and appropriate staff development for teachers, as well as consultant help for them on a more individual basis. The need for this assistance *must* be appropriately addressed if innovation is to succeed.

Lack of leadership. The key to overcoming the barriers identified here is sufficient quality leadership. The lack of such leadership becomes itself a barrier. The importance of this leadership deserves the most careful consideration. Research has established its critical importance. The developing of leadership capability deserves the highest priority on the part of any decision makers in a school setting where change is being sought.

A non-nurturing environment. Innovation requires a supportive climate for those leaders seeking to bring about change.

School officials at the highest level need to give this matter significant attention if they want others in the system to be innovators.

Possible Pitfalls

The following pitfalls are examples of the errors or misguided efforts we have seen schools or school districts make in an attempt to improve or reform their local science program. Although not exhaustive, these examples illustrate the complexity of making the changes advocated in the earlier chapters.

Pitfall: Questionnaires are Not the Same as a Needs Assessment

To many people a needs assessments is synonymous with doing a survey. Questionnaires administered to parents, teachers, administrators, and often students represent their current knowledge and feeling at the time of the questionnaire. Knowing how people feel about a proposed change or the current status of program may be very useful information in planning strategies to bring about the change, but should not represent an assessment of the "need." A needs assessment should represent a direction or need in the future, not a measure of the status quo. Such procedures have been outlined in Chapter 3 and involve literature studies, delphi procedures, brainstorming, task forces, consultations with experts, etc.

Pitfall: A Strongly Supported Need Does Not Mean Immediate "Buy-In"

Often, needs assessments are done with an attempt to measure the degree of "buy in" of an audience, typically teachers, for a new program or other innovation. The hope is that if a positive result is obtained from the needs assessment, implementation of the new program will proceed in a fairly simple and orderly fashion. It is the authors' experience that few, if any, significant innovations are made without a major amount of effort, and possibly even pain. As one example, it is not unusual for teachers to respond in a questionnaire that they believe their elementary science program should contain a greater amount of "hands on" activity or that the high school chemistry course should include more time spent on societal and environmental issues. However, acceptance and implementation of such a program may be quite another matter. The implementation of the hands-on program requires extensive new work on the part of the teachers, provides far less "coverage" of material, and even creates a mess in the classroom. Most of these consequences could not be foreseen by many teachers responding to the needs assessment questionnaire, but when faced with the reality in their classrooms they may find it more difficult to implement than they originally envisioned.

The high school chemistry teacher who begins to add topics on air quality, water pollution, medical impact of biochemical discoveries, etc., may find that there is less time for the conventional topics normally covered in a chemistry course or that fellow science teachers in the department look with disfavor upon this choice of new content. There may even be a drop in the standardized test scores at the end of the year. All of these results may have been unforeseen by most teachers completing the earlier questionnaire.

The Stages of Concern (SoC) outlined in the CBAM Model are not understood by most teachers and administrators, and the problems and concerns that they encounter during the implementation process cannot be foreseen in the results of a questionnaire. The Stage of Concern Questionnaire itself can be a type of needs assessment used to identify the current concerns of the participants involved in the innovation and to design the assistance needed to improve the quality of the implementation. This particular questionnaire does have some predictive qualities since research has shown that all people move through essentially the same stages in the same order—albeit at a different rate, depending on the innovation, their experience, personality etc. Most other locally developed needs assessment questionnaires do not have this predictive quality since they are designed only to measure reactions of the participants at the time the response are indicated.

Pitfall: Improving the Quantity Does Not Improve the Quality or More Is Not Necessarily Better

A reading of some of the national reports seems to indicate that the major problem of American education—or science education in particular—is that there is not enough of it (quantity). The school day or year needs to be longer, and more needs to be covered. Other reports indicate that the objectives of yesterday are not appropriate for students' needs today and tomorrow; or that students today are not being motivated by the current pure science objectives; or that current curricula "cover" a lot of detail without producing student understanding of important fundamental concepts. More of the same probably is not the answer. The question of what ought to be taught (quality and appropriateness) must be addressed first.

Increasing the amount of time spent on science may improve the achievement of students if achievement is based upon the number of items correct on a standardized test or the number of objectives completed but it does not indicate that the quality of instruction or the quality of student experience has improved.

Changing the appropriateness of science goals, e.g., from pure science to STS, does not necessarily change the quality of the experience on the part of the student. There is the potential that students will be more motivated by "relevant" science, but a teacher who continues to present science in a dry lecture oriented format, even with the new materials or objectives, is not likely to attract and motivate many more students.

Quality is a somewhat difficult entity to describe, but is generally used to describe the effectiveness of the instructional strategies used by the teacher. An improvement in quality of instruction certainly should impact how well a student learns what has been presented, it may not increase the quantity to any great degree. There is some evidence to suggest that it might reduce the number of objectives covered since quality instruction usually means taking more time to go in-depth for more complete understanding on a fewer number of topics. Extensive laboratory work, a good follow up discussion, and a review to check for understanding all take time, but evidence indicates that they do produce a higher degree of learning of the material that is presented in this fashion.

As science educators we should be interested in improving all three dimensions: quality, quantity, and appropriateness; all have something significant to offer, but budgetary and other resource constraints may restrict us to only work on one type of improvement at a time. If this is the case, we should not delude ourselves into believing we have affected the other two.

Pitfall: Developing or Adopting a New Program Will Not Necessarily Improve Science Teaching, or Meaningful Change Takes More Than a New Guide

The first form of this pitfall is to develop a new set of goals and objectives, called a syllabus or framework, and expect that this document will improve the science teaching in the classrooms, a district or a state. Such documents are usually not intended by their developers to be a complete end in themselves. Their usual purpose is to define the curriculum for a state or district and provide the basis for further improvement efforts, such as the development of more complete teaching guides, staff development programs and a student assessment or evaluation program by local schools. Unfortunately, some districts and states have lost sight of this purpose and see the syllabus or framework as an end in itself. Until a program has been translated into a change in the behavior and the beliefs of the teachers involved, no change or improvement will occur.

Sometimes this pitfall is carried one step further and a complete, well-developed guide is written, delivered to schools, but little or no support is provided in the form of staff development, follow up consultation, or even equipment and materials. There is little evidence that any significant number of teachers can change their behavior to any large degree based upon only receiving and reading a new guide.

Pitfall: A Local Committee Can do the Best Job

Another pitfall often encountered is to give the job to a committee of representative teachers to rewrite a curriculum without outside input. Much like the questionnaire or needs assessment described above, this process will simply describe the current knowledge and belief of the committee and probably will not move the district or school to a new and improved program. Teachers certainly must be involved, and may well take the lead in important ways, but new ideas and methods are needed from the outside. Often the innovative impetus can come from the inside if a small group of innovative teachers are provided with resources and support to begin the change on a small pilot basis and then supported in their efforts to transplant it to the rest of the school or district. But if a committee is simply given the assignment of developing a curriculum without strong innovative leadership, or explicit reference, or guidance from outside projects, the committee will usually develop a program that is centered around the norm of the existing group. Such a procedure often provides a certain amount of "buy in" or acceptance early on and keeps peace among the faculty, but rarely results in the development and implementation of a major improvement.

Pitfall: Implementation Efforts Are Not Robust Enough, or This is the Year for Science; Next Year is for Social Science

A common pitfall is to put an extensive amount of effort into the development and even pilot testing of materials and then put only a limited amount of time and effort into the implementation. This pitfall is actually an extension of the one described above concerning the development of materials without implementation activity. Unless the implementation support activities are viewed as a way to help teachers change and improve their teaching behavior, the improvement effort will probably fall short of its goal.

Although there is no clear evidence as to exactly how much staff development is required, or over what period of time it should be presented, the evidence from the CBAM project (Hall and Pratt, 1984), DESSI study (Crandall, 1982) and others indicates that a minimum of a year (and probably longer) is necessary. Experience by one of the authors in conjunction with the CBAM project indicated that three full days of staff development spread out over a year was able to move a large number of teachers to a largely mechanical level of use of a new hands on elementary program. Still more time was necessary to get much of the population to a higher, more comfortable routine level, in which they felt enough at ease to continue with the program without extensive support. One outcome of the DESSI study was that both early support and pressure as well as later, ongoing support and pressure by the instructional leaders were required for the program to be successfully implemented and later institutionalized.

In some districts where the resources available for curriculum development and implementation are limited, a teacher or staff development person is given a year to develop and implement a new science program at either the elementary or secondary level. The following year the person and resources are then shifted to the next area of emphasis, such as social studies, language arts, etc. Such an administrative maneuver does not allow sufficient time for the program to be developed, and far from sufficient time for the program to be adequately implemented. At the elementary level this is compounded by the fact that during the second year teachers must shift their attention from the first area of innovation to the second. This compounds their frustration and may result in teachers rejecting all of the new programs.

Pitfall: Staff Development Programs That Only Tell Teachers What To Do

Another pitfall is a staff development program that only describes what the teacher is to do in the classroom and does not demonstrate it and provide for practice. If the program is indeed a hands-on program for students, staff development for teachers should also be hands-on. It hardly seems necessary to identify this pitfall, but it is amazing how many times we hear of teachers being talked to at an in-service. Adults need the opportunity to learn how to teach by constructing their own understanding from their experiences. Hearing someone else describe an activity or experiment is probably one of the least effective experiences from which to construct a new and improved way of teaching science.

Pitfall: Too Little Collaboration

The power of collaboration may be the most overlooked strategy in the school reform effort. We are talking about collaboration on many fronts: between teachers and administrators, between the school staff and parents, but most importantly among teachers. The isolation among teachers seems to be a major deterrent to change, yet many improvement activities do not include opportunities for teacher-teacher or teacher-administrator interaction in the improvement plan. Even if all other elements; such as awareness, materials acquisition and resupply, staff development, student assessment, and program evaluation, have been included in the plan, collaboration is still critically important.

One of the most compelling reasons to consider collaboration is that it works. Successful change efforts seem to include plans and arrangements for people to work together. Schools are cultures with their own mores, traditions, and other mechanisms to maintain the status quo and resist change. Although a few organizations and schools accept unique, innovative and different behaviors by individuals, this usually is not the case; the members usually move through a change together as a group. One process to overcome this resistance is to make it socially acceptable to change. Collaboration enhances and facilitates this process of change or evolution.

Add to the social resistance to change the fact that changes in teaching behavior (or any behavior) require new skills and knowledge. The skill part is almost impossible to acquire in a staff development setting outside the classroom itself (Joyce, et al., 1989). Therefore, an effective staff development strategy is the practice of new behaviors in teachers' own classrooms, followed by sharing sessions among a team or department working on an innovation. This often occurs spontaneously if a group of teachers are a part of a team or department that is working together, but it can be formalized and structured with mutual observations and peer coaching (see Chapter 5) by the teachers and administrators involved. Collaboration provides a way of both developing and practicing the new skills and making them socially acceptable in the local school culture. As indicated in Chapter 6, teachers need a context in which they mutually adapt the innovative practice by creating their own individual and collective understanding of it; a process critically necessary for successful implementation.

The barriers and pitfalls are numerous, but with forewarning, as well as sufficient leadership and resources, they can be overcome and real improvement achieved. The importance of quality science education for American youth is great. Given a long term commitment at the local level it can be improved. Good luck in translating such commitment into results.

References

Crandall, D., et.al. (1982) *People, Policies and Practices: Examining the Change of School Improvement*. Andover, MA: The NETWORK, Inc.

Joyce, B., Murphy, C., Showers, B., & Murphy, J. (1989) School renewal as cultural change. *Educational Leadership*. *47*(3)

Hall, G. & Pratt, H., There really can be a symbiotic relationship between researchers and practitioners: the marriage of a national R & D Center and a large school district. Paper presented at the meeting of the American Educational Research Association

Epilogue

In the scenarios that opened this book, educational leaders were facing the apparent need to respond to deficiencies in their science programs. When they set out to address this need they conceivably could have landed in one of the pitfalls just described in the last chapter, or for some other reason failed to get the marked improvement they expected. On the other hand, they could have avoided the pitfalls and initiated a process that produced the many positive outcomes one would expect from a systemic endeavor that is "done right."

To portray these positive outcomes and illustrate the desired results of the approaches advocated in this book, two documents are presented here: the executive summary of an evaluation report and a letter written by a science department chairperson.

A District-wide Improvement Process

The first of these two items—the executive summary of an evaluation report—is the work of an evaluator who is reporting to a school board on the evaluation of a new 7th grade life science course, which its district has introduced. Of particular interest is that the report addresses the process used to initiate the new course, as well as the course itself; it is presented below.

Executive Summary

Summative Evaluation of New Seventh Grade Life Science Program

Prepared for
School Board of Buffalo County School District
Timber Mountain, Colorado
By
Dr. I. M. Richter

This report constitutes the summative evaluation of a four-year endeavor for revamping the seventh grade life science program in the Buffalo County Schools. In response to the original charge given to the evaluator by the board four years ago, this report addresses the *process* by which the new program was introduced, as well as the results of the program itself.

The process began with a needs assessment that included a review of national reports on what was needed in middle school life science programs, and a participatory process in which staff and representative parents reached consensus on the focus of a program for the district. The process took considerable time, but the foundation laid by this process was evident in subsequent years, and it is recommended that the administration use a similar process in its future curricular change endeavors.

The program selected, *Life Science: A Human Approach*, was chosen because of the overlap of its content (about 80%) with the district focus and the compatibility of its instructional approaches with district preferences. A short guide for the program was developed locally and identifies the units of highest importance in terms of district goals and those that are strong candidates for elimination. Considerable room for individual teacher choice remains, however, and teachers were encouraged through the guide and in-service classes to limit the number of units and teach them thoroughly, rather than attempt to cover all the units in the program.

The process of putting the program in place extended over three years and included in-service education classes specifically designed for introducing this program, regular monitoring of teacher concerns throughout in accordance with the Concerns Based Adoption Model, assessment of the extent to which the program was used as intended with an observation instrument based on a profile of the ideal program, and assistance to teachers from a cadre of selected teachers. Assessment of this approach indicates that it was largely a success although better communication probably would have been desirable. Interviews conducted as part of the assessment indicated that most teachers feel the process was responsive to their concerns, individual assistance was available when needed, the program's demands on their time are reasonable, and the focus of the program is consistent with the original goals identified for life science in the district.

The primary means of assessing the outcomes of the program was an end-of-year test developed specifically for use in the district. It was limited to approximately half of the units in the program (those designated "highly recommended" in the guide) in order to encourage the "less is more" philosophy and allow some teacher choice of units. Other forms of assessment by teachers were encouraged in the teacher in-service classes. Student outcomes with respect to the various program objectives are contained in this report. The data indicate favorable results for the program, as compared to the old program it replaces. Profiles are provided which indicate a better match of curriculum goals and actual instruction under the new program than under the old one. With two exceptions, the level of attainment of the several goal areas is at a "moderately high" level. In terms of student outcomes, the program is judged to be successful.

The process by which the program was introduced also is judged to be successful and worth the resources it required. Data from teacher interviews and questionnaires indicate that they think—compared to previous district curricular implementation efforts—they received more assistance, their various concerns were heeded adequately, and the program is being used in a manner closer to what was intended. Both the specific costs and benefits of the process are described. Although the full report identifies such issues as the relationship between district-wide implementation efforts and individual school decision-making which must be addressed more fully in future curriculum implementation efforts, the overall assessment of the process is that it was successful and deserves continued use.

A Science Department Improvement Process

The second document presented here is a letter written by the chair of a senior high school science department about a three year effort to revamp their science program, a process prompted initially by release of a new science curriculum guide by the State Department of Education. The letter is written to a fellow science department chairperson in a neighboring district—and at one time a fellow teacher in their school—who sees a need to do something similar. The letter addresses the highpoints of the process and is expected to be the basis of further discussions they plan to have in the future.

Dear Joel:

I promised to write you a letter about our science revision at Rock Ridge High School, so you would have a summary in hand before we meet to discuss some of the details. Being so close to the process, it is hard for me to pick out the most important points, but I will do my best.

It all started when the state put out the new science curriculum guide, probably as a result of the ideas coming out of the SS&C Project and Project 2061. When they released the new guide and also announced that there would be about $2000 a year for three years available to schools wanting to revamp their program, I decided to go for it. The teachers in the department agreed to go along with the idea, and in fact, except for Mark and Joan it was more than just going along with the idea; they were quite enthusiastic. We applied and got the grant; we also started into a process that was more work than we had imagined.

You probably are wondering just what has made this thing go. I'll have to admit that to a large extent it was me. Not that everyone hasn't been working hard; they have. But I take my job as chair seriously; so, I worked hard at getting us organized, planning meetings and calling around to take care of problems and get things we needed. Our principal, Charlie, has been very supportive, but it has been mostly moral support. That's not really a criticism. It seems that the approach in all the schools in our district is one that is very decentralized, right down to the department level. He has helped in getting some resources a couple times, but he never sits in on our department meetings or anything like that. He says good things about us to the other departments and down at the head shed, so we feel good about him. Keith, our district science supervisor has been super about giving ideas, encouraging us and tracking down materials from other places we couldn't find, but as you understand, in a district with 73 schools of various levels, he couldn't be here to make the process go.

By the way, we did not get a lot of money to do this. The $2000/year from the state mostly went to pay for some summer salary time for our department, but it sure didn't go far. Charlie also lined up some district money for us that made it possible, with the state money, for seven of the eight of us in the department to get paid for one week of work each of the last three summers. Other than that, we haven't had any extra money of any consequence for doing what we are doing. Of course, a lot of extra time has come out of our hides.

So, what did we actually change around here? The big thing, of course, was totally reorganizing the curriculum. We now have Integrated Science I, Integrated Science II and Integrated Science III instead of biology, chemistry and physics. The new courses are mostly, but not entirely, a reorganization of the content of the three courses they replace. In fact, when I think about it, I realize it is considerably more than a simple reorganization. We have left things out, so we could make other changes. We added in some earth science that wasn't there before, and we definitely give more attention to the applications of science. A good illustration is that our chemistry is based on the *ChemCom* materials; we have a full classroom set of *ChemCom* books in each room for students to use in class. Each year students study some of all of the science subjects.

The other big thing the kids see is that we do lots of labs—far more than we used to do. And most of these labs are really open-ended; we have searched all over the place to find ones that are really good. I guess there is one more thing the kids notice—we are using portfolio assessment. It has been hard for some of us to really get going with it, but by-and-large it is working, and I don't think we want to go back to our old way of assessing.

In all of this we are trying to teach in a different way, but I don't know if I can really explain in the abstract what it is. I have heard some people talk about a constructivist approach to teaching, and that may be what we are trying to do. We definitely are trying to teach the more fundamental concepts in depth and not get caught up in vocabulary and detailed facts. "More is less" is the phrase we sometimes use to explain some of what we are doing.

Well, I imagine you are also wondering about how we actually have managed to pull off all of this. Simple—LOTS of hard work. But we are not complaining; I think all of us would be willing to do it again—except maybe Mark and Joan. They are into it, but if they had to do it over again, I am not sure.

I think the secret to our willingness to work so hard and get good things done is that *we work together*. Our teams for each level meet every week for an hour after school to plan. In addition, we depend on each other for help. After all, each of us has a major in one of the sciences and only a couple college courses in the others, so when we teach Integrated Science we have some digging to do. We catch each other between classes and in the prep rooms frequently to get an explanation of some concept or how to operate a piece of equipment. If there is any secret to our success, that's it— working together.

As you can imagine, we have had some difficulties as well. The biggest one is that we have put the course together from parts of various books we keep in the classroom. Because each class is using several books we don't have enough that the kids can take them home on a regular basis. We keep them in the room so we are sure we always have enough for everyone when we use them in class. I hear that there are some books coming out that are closer to what we want, so you probably wouldn't have to face this same problem if you want to do something of this sort. The danger is that you might not do the kind of planning that really helped us to change our teaching. One thing I think you should do even if you have a better book situation is to collect your own set of labs and teaching activities as a team. And most importantly, don't fall into the old trap of having the kids read the book and then talk about the reading in class. That's the deadly old crap we have been working so hard to get rid of.

I've been talking as if what we are doing is a success. We really believe it is—even Mark and Joan. I suppose you are wondering what I base that on. It's the results with the kids. Some people want to know about test scores, so we took the portions of a standardized science test that fit our curriculum and compared the results with students in sections of our old program that we were running in parallel during these first three years as we developed the new program. In a nutshell, the results are essentially the same. It is the other results that impress us. The kids like science more! More kids are taking science beyond the minimum number of courses. And we think that our portfolio assessments tell us that kids are learning more about the big ideas of science and their applications.

There is one more result that is very important—we like it better. A lot of it is that the kids are more enthusiastic about science, but I think it is something more. We are convinced that what we are teaching is more important and relevant. Somehow it gives us more satisfaction.

In a nutshell, that's it. When we get together I can fill in more of the details. My advice—assuming you are asking for it—is that if you and the other teachers are inclined to head in this direction, go for it. The key is working together. Of course, leadership is important too, but I think you are the kind of person who has the initiative to get things organized and also encourage people to work together.

See you soon.

Sincerely,

Sarah

INDEX